Studies in Systems, Decision and Control

Volume 219

Series Editor

Janusz Kacprzyk, Systems Research Institute, Polish Academy of Sciences, Warsaw, Poland

The series "Studies in Systems, Decision and Control" (SSDC) covers both new developments and advances, as well as the state of the art, in the various areas of broadly perceived systems, decision making and control–quickly, up to date and with a high quality. The intent is to cover the theory, applications, and perspectives on the state of the art and future developments relevant to systems, decision making, control, complex processes and related areas, as embedded in the fields of engineering, computer science, physics, economics, social and life sciences, as well as the paradigms and methodologies behind them. The series contains monographs, textbooks, lecture notes and edited volumes in systems, decision making and control spanning the areas of Cyber-Physical Systems, Autonomous Systems, Sensor Networks, Control Systems, Energy Systems, Automotive Systems, Biological Systems, Vehicular Networking and Connected Vehicles, Aerospace Systems, Automation, Manufacturing, Smart Grids, Nonlinear Systems, Power Systems, Robotics, Social Systems, Economic Systems and other. Of particular value to both the contributors and the readership are the short publication timeframe and the world-wide distribution and exposure which enable both a wide and rapid dissemination of research output.

Indexed by SCOPUS, DBLP, WTI Frankfurt eG, zbMATH, SCImago.

All books published in the series are submitted for consideration in Web of Science.

Anton Gerunov

Risk Analysis for the Digital Age

Springer

Anton Gerunov
Faculty of Economics and Business
Administration
Sofia University "St. Kliment Ohridski"
Sofia, Bulgaria

ISSN 2198-4182 ISSN 2198-4190 (electronic)
Studies in Systems, Decision and Control
ISBN 978-3-031-18099-6 ISBN 978-3-031-18100-9 (eBook)
https://doi.org/10.1007/978-3-031-18100-9

This Springer imprint is published by the registered company Springer Nature Switzerland AG
The registered company address is: Gewerbestrasse 11, 6330 Cham, Switzerland

Introduction

The new digital economy is the reality we all live in. The last decades have seen the exponential growth of advanced technologies that have led to a significant rise in the automation and digitalization of all business processes. So goes the trivial observation any pundit can quickly make. This is, however, merely the tip of the iceberg. The enabling factors of this transformation, the driving undercurrents, are those fundamental developments that change not merely the surface but the underlying landscape of economic and social activity. We are observing a new type of industrialization which is driven by information assets such as datasets, advanced digital infrastructures, extensive codebases, and trained Artificial Intelligence models. From an economic standpoint, this leads to a decreased marginal cost of production, decreasing share of labor in the final product, increased personalization of products and services, and the concomitant increase in the complexity and uncertainty of business activity.

What is more, qualitatively new phenomena are now taking center stage in the global economy such as social networking, smart devices, and an explosion of data that power scaringly accurate machine learning models for classification and prediction. These novel and markedly digital phenomena are hardly limited by physical boundaries and cannot be expected to behave the way known ones do. The digital world is an exponential winner-takes-all type of environment—something that can be observed in the way engagement, impact, and revenues behave. It is also a detailed world—sophisticated technology enables microtransactions, micropayments, and granular behavior monitoring and influence. Finally, it is a relatively democratic world—barring pertinent limitations of skill—it takes very little to become a full-fledged participant, producer, or even a threat.

All this creates innumerable opportunities, but also dramatically expands the risks individuals, organizations, and nations face. The risk universe is not only vastly larger but more complex to comprehend and analyze with traditional tools. This book aims to bridge this gap by providing initial ideas on how to approach novel risks and carry over traditional tools and approaches to the new digital reality. The discussions presented here are merely starting points in a long journey and it is expected that they will bring questions and controversy, but also attract healthy criticism. The end goal is clear—start the discussion that culminates in a rigorous and reliable methodology for understanding and managing risk in a world that increasingly blurs the boundaries between real and virtual.

Chapter 1 of the book starts the journey. It begins by outlining the fundamentals of risk and setting the stage with appropriate definitions. The risk curve is presented and also an initial pass at modeling risk by leveraging qualitative evaluations and prioritization within the framework of a risk matrix. Those intuitions are then formalized by taking recourse to classical utility functions and formally deriving risk preferences as defined in the Arrow-Pratt measure. This measure gives an idea of how to measure risk with the risk premium—the reward one received from bearing uncertainty. The chapter then looks at empirical data which is used to estimate the risk premia in 26 European countries over a period of 15 years, ending in 2020. This simple metric may not be enough to capture a complex process that changes its dynamics over time. This is why we suggest to move away from simple estimates and instead measure the risk premia for specific types of periods (such as risk during boom and risk during recession). The proposed way to do this in practice is by using a Markov switching model that estimates the optimum number of states and their relevant parameters. Regime-variant risk premia for those 26 European stock markets are calculated and discussed.

Chapter 2 introduces the reader to more standard risk metrics. It takes the usual route of treating the risky variable as a random one that can be described by its statistical distribution and outlines a few popular distributions that are in use in risk analysis—the Normal and Cauchy ones, the Student T, the Exponential, and the Lognormal one. Investigating those, one is compelled to reach the conclusion that pertinent risks lie in the tail of the distribution and may be modeled by calculating the Value at Risk (VaR) and the Expected Tail Loss (ETL). From a practical standpoint, there are many ways to do this, taking into account different periods and relying on alternative assumptions. The chapter presents actual data on 25 European stock markets for a period of 16 years and calculates their respective VaRs and ETLs under different assumptions. We show that the quantitative difference in estimates may be very large under different assumptions, reaching 42% for VaR and over 100% for ETL. We further show that judging by the forecast error, the best-performing estimation approach is to use non-parametric estimates. At any rate, regardless of assumptions, both VaR and ETL fail to account for the catastrophic rise in risk in 2020 engendered by a global health pandemic.

Chapter 3 switches its focus to digital assets. It defines information assets and digital financial assets and presents an overview of the major types in both classes. Risk estimation for information assets may have to include a formal valuation, and some approaches to do this are proposed. On the other hand, digital financial assets may be traded in a market, but their behavior varies significantly from those of traditional financial assets, particularly in the early stages of their introduction. The chapter presents a detailed study of a major new emerging digital asset class—the cryptocurrencies—by reviewing a set of 20 coins with large market capitalization. Traditional risk metrics are calculated but they are shown to be missing a part of the picture. We thus propose enhancing the classical mean-variance optimization with a new major risk typical of cryptos—liquidity risk. Essentially, this means including a measure of trade volume in the risk optimization decision, thus moving it away from the familiar two-dimensional optimization space and into three-dimensional one. An example of this new concept is presented, underlining that it serves as an extension to traditional theory.

Chapter 4 takes recourse to a markedly digital theme—the rise of networking. The emergence of the network society and network economy, powered by information technology, has also transformed risk. Uncertainty is now inherent in the network structure of interactions and risk management may benefit from taking a more detailed look at it. This chapter briefly presents the fundamentals of network theory and some relevant applications in Social Network Analysis. Key network parameters such as centrality and clustering are presented, and crucial network processes such as diffusion are conceptualized. To motivate the discussion, we present both a theoretical model of cascading network failures that results in a financial meltdown, as well as a Ponzi-scheme network. Understanding risk as a network phenomenon also necessitates the uptake of new tools. We propose the Bayesian Belief Network as a useful tool for modeling networks where connections between nodes are unknown. The chapter uses statistical data to recreate the network structure of non-diversifiable macroeconomic risk that affects the digital sectors of the European economy. Risk betas are calculated and interpreted.

Chapter 5 takes on the Holy Grail of risk management—analyzing and predicting rare and potentially catastrophic risks (so-called "black swans"). The main approaches to do that are presented, covering expert judgment, statistical methods, prediction markets, and simulations. The chapter reviews the potential benefits but also the challenges in using all of them. The key challenge invariantly remains that the analyst has either very limited or practically no data on those rare risks. We propose to surmount this by leveraging a rigorous and well-thought Monte Carlo simulation with extensive sensitivity analysis. Such a simulation is created to model very rare risks that affect the sector of information processing services in Europe as a proxy of the overall digital economy. The Monte Carlo simulation is run over 10,000 iterations under four alternative scenarios, and the results are commented on in depth. The main conclusion is that digital phenomena seem to be better modeled by non-normal distributions that provide for longer tails such as the exponential and the lognormal ones.

Chapter 6 investigates how individuals perceive risk and how their behavior changes in a digital environment. It begins with a plain vanilla presentation of main theories on individual risk perceptions starting off with classical utility theory, going through some of its many modifications (and their respective utility functions and risk measures), and into the realm of behavioral economics. The chapter then presents recent advances in neuroeconomics that show how the structure, size, and composition of the brain affect risk perceptions. Crucially, it proceeds with how the online environment mediates human perception, engendering whole new sets of heuristics and biases online users fall prey to: reputation, endorsement, consistency, expectancy violation, and pervasive intent heuristics. An original economic experiment is then

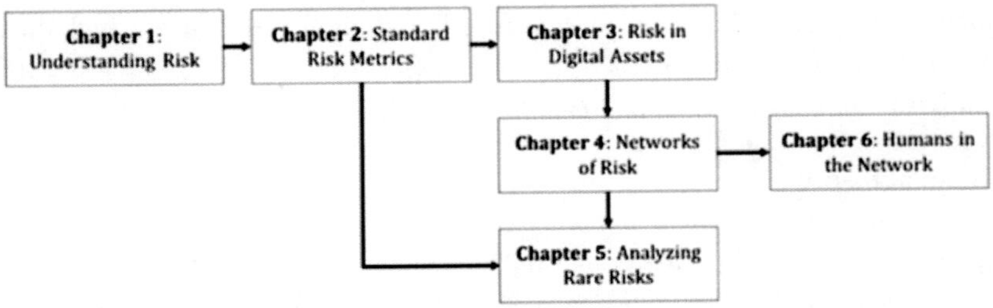

Fig. A Overall Structure of the Book and Chapter Connections.

presented that investigates how subjects perceive risk and make decisions in a simulated online social network as opposed to a control group with no network access. The resulting decisions tend to be less optimal in terms of outcomes whenever a proto-network is present. Furthermore, we investigate the ability to forecast individual decisions using 55 leading machine learning algorithms and find out that the presence of a social network decreases predictability.

These chapters can be read either sequentially or ordered at the reader's pleasure. Still, in some cases preceding chapters give more background and context to the following ones. This is illustrated in the following *Figure A*, whereby arrows indicate useful background from one chapter to the next. Such routes through the book may also be useful for undergraduate or graduate courses where the instructor may wish to use only parts of the book.

In addition to the problems and cases presented in this book, the avid reader will surely think of many more that are relevant to analyzing risk in a digital environment. We hope that these can also be tackled with the methods and extensions proposed here. The chapters of the book have likewise relied on the extensive and sometimes multidisciplinary research conducted in the fields of risk management, decision science, behavioral economics, finance, digital strategy, and network science. This work also has a number of original contributions that may be of interest to academics and practitioners:

- Calculates long-term risk premia for 26 European markets over a period of one and a half decades.
- Presents regime-variant risk premia for 26 European markets, and proposed using a Markov switching model for estimating regime-variant risk for digital phenomena.
- Compares and contrasts VaR and ETL estimates under different assumptions, showing that non-parametric calculations exhibit the best performance.
- Proposes a novel approach for analyzing cryptocurrency risks that includes standardized volume traded as a proxy for liquidity risk.
- Demonstrates the application of the new cryptocurrency risk analysis approach by moving from the traditional 2D mean-variance optimization space towards a 3D space that includes standardized trade volume.

- Creates a rigorous network-driven risk model for the digital sectors of the European economy—Codes J62 (Computer programming, consultancy, and related activities) and J63 (Information service activities).
- Estimates risk betas as contributions to conditional density in the Bayesian networks for the digital sectors in the European economy.
- Estimates and compares four Monte Carlo models for studying information service activities revenue, thus underlining the necessity of using non-normal distributions when modeling risk in the digital economy.
- Conducted (with co-authors) and analyzes an economic experiment showing the significant effects of social networking on individual risk perceptions and decision-making.
- Tests 55 alternative machine learning algorithms for predicting individual choice and outlines the best-performing ones.
- Using the results from the decision prediction exercise, we show that exposure to the social network makes the decision more erratic and less prone to prediction.

The journey towards understanding digital risk is a long but exciting one, and its importance will only grow as the speed of digital transformation accelerates. This book is probably a little step in this march, but the author hopes that reading it will be as enjoyable as writing it.

Contents

Chapter 1
Understanding Risk

1.1 Introduction

Risk and uncertainty are fundamental aspects of economic and social activity.[1] The challenge of making reasonable, if not strictly rational, forward-looking decisions under uncertainty is aptly summarized by Niels Bohr who stated that "prediction is very difficult, especially about the future".[2] And yet, predict we must in order to act. This practical imperative gives rise to the science of studying risk that focuses on how one can manage the possible future eventualities in order to achieve a desired outcome. Despite the striking simplicity of this definition, risk management relies heavily on randomness and probability and their interpretation reflects a deep underlying view of how the world functions. Essentially, how one views the world defines how one perceives risk and acts upon it.

The challenge for understanding and taming risk, on the other hand, has never been bigger. The growing complexity of the modern economy and society has spawned a dramatic increase in the risk exposure of both individuals and organizations.[3] The methods and approaches to managing this new complexity have not necessarily kept apace with those rapid developments and sometimes one finds them to be wanting. Key sectors of the economy such as finance, insurance, critical infrastructure, and medical research tend to be the flagships of implementing sound risk management but many more lag behind. In addition to that, a number of crucial but difficult to measure or intangible risks are managed either poorly, or not at all. This has led some observers to proclaim that risk management is broken and in need of fixing.[4]

This book takes a different perspective on this issue: it recognizes the complexity of aligning subjective individual risk perceptions with the objectively existing randomness in a complex and adaptive systems with emergent properties such as

[1] See e.g. Bullen (12).

[2] As cited in Ulam [73], p. 286.

[3] Chernobai (15).

[4] Hubbard [41].

A. Gerunov, *Risk Analysis for the Digital Age*, Studies in Systems, Decision and Control 219, https://doi.org/10.1007/978-3-031-18100-9_1

the national economy. Thus, it considers the science of risk as a work in progress that moves towards reaching a certain maturity that will both aid practical activities in finance, medicine, technology and numerous other fields, as well as provide a deeper understanding of our world. The first step towards this is understanding risk, how it is defined, what are the different types of risks and how it can be modeled formally in a rigorous mathematical framework. This chapter intends to do exactly that, and also to contribute a few empirical facts about risk perceptions and premia in developed and emerging markets in Europe.

1.2 Definitions and Typologies

1.2.1 Risk: A Short History

The fundamental concepts of risk and uncertainty have always been in existence—the world has always been characterized by randomness from the quantum level all the way up to human interactions. Economic activity has naturally been largely exposed to this from the uncertain agricultural yields to the looming possibility of war or natural disaster, but for long the planning for these events has been mostly implicit or at least non-rigorous. Covello and Mumpower[5] date back the earliest risk management consultants to 3200 BC—it was a small group called the Asipu that lived in the Tigris-Euphrates valley. When enlisted to make a determination about a risky prospect, a member of the Asipu would collate alternatives, weigh their outcomes, prepare a ledger that balances them out and even hand out a final report to the client on a clay tablet.[6] The Asipu's conclusions were grounded in authority, rather than in probability. Indeed, over the next centuries before the formal study of probability, risk management was intuitive rather than formal. This led to four generic management strategies: buy some form of insurance, rely on government intervention, set out rules in common law, or depend on self-regulation.[7]

These approaches merely provide possible buffers against the outcome but fail to distinguish how likely it is, and thus what is the optimal size of the buffer and the relative importance of different risks. This all changed with the rapid development of probability theory from the sixteenth century onward. Among the first to apply probability theory to individual decisions under risk is Daniel Bernoulli, and he was also among the first to note that individuals do not make rational determinations when faced with such prospects—a finding that came down as the St. Petersburg Paradox. A number of other paradoxes ensued, with scientists showing sometimes large dissonance between objective outcomes and subjective perceptions, driven mostly by a divergence between objective and subjective probability. It was

[5] See Covello and Mumpower [24].

[6] See also Grier [37].

[7] See again Covello and Mumpower [24].

the work of early scholars such as Bayes and Bernoulli that marked the beginning of formally modeling uncertainty (i.e. probability) and formally modeling individual preferences (i.e. utility) as a major way of looking at risks and risk management.

It is widely considered that the founding father of the study of economic risk in the modern sense is Frank Knight.[8] In the early twentieth century he distinguished between four different economic situations depending on the degree of randomness they are subject to: these ranged from no randomness at all to absolute randomness leading to impenetrable outcomes. Some researchers put the start of the formal study of risk management even later—during and right after the Second World War.[9] At any rate, the last decades have seen a large interest and rapidly expanding research in the field of risk. As a rule, the financial sector has been a critical laboratory for methods and approaches to managing risk but other fields such as medical research and mission-critical infrastructure (e.g. nuclear reactors) have developed a rich and sophisticated tradition of their own.[10] The recent rapid developments in information technology and the associated explosion in the organizational exposure to cyber risks has also seen a swiftly growing interest in the study of information and data risks.[11] This barely changes the fact that the field of risk management remains relatively young and many challenges for measuring and analyzing the risks that truly matter remain unsolved.

1.2.2 Defining Risk

As economists were setting their eyes upon decisions under uncertainty, the precise definition of this concept was becoming increasingly important. One of the early and very influential definitions was given by Frank Knight.[12] He surmised that any given decision situation could be characterized by the its outcomes, and the likelihood of each outcome. Thus we can denote the outcome, or result, of a given decision by x_i, and its associated probability as p_i. In a well-defined decision problem we should be able to completely cover all possible outcomes, thus imposing the condition that $\Sigma p_i = 1$. All possible economic situations according to Knight belong to one of the following four types:

- **Certainty**—the outcome(s) are clear and well understood by the decision-maker and they materialize for sure, i.e. with a probability of $p = 1$. Such situations are generally within the scope of physical laws—e.g. water boils at 100 °C at sea level, or falling object accelerates with a constant velocity of about 9.8 m/s^2. In economics, it is the exception that a decision-maker is able to predict with

[8] Knight [52].

[9] Dionne [30].

[10] For a critical review of these approaches, the reader may consult Hubbard [41].

[11] Egan et al. [31], Evans [32].

[12] See Knight [52]; but also Keynes [49].

complete certainty what the outcomes are. This is due to a number of reasons ranging from the objective riskiness of the economic environment[13] through bounded rationality,[14] all the way to a large number of behavioral biases that people exhibit.[15] Decisions under complete certainty are thus very rare, and their handling is mostly trivial. Even in more sophisticated settings such as fully rational consumer choice under certainty, the outcome can be easily determined through rather straightforward optimization.

- **Risk**—this is a situation in which the decision-maker can identify both the complete set of outcomes x_i, and their respective probabilities, p_i. The archetypical example that has bored many a student of mathematics and statistics is the fair coin toss—it can land either heads or tails, and each of the sides comes with an equal probability of $p = 0.50$. This example, in fact, has irritated some so much that Nassim Taleb uses it a warning tale against blindly trusting pre-defined probabilities in the face of mounting evidence to the contrary.[16] Apart from this simple example, some more complex phenomena such as economic growth, credit repayment, fraud, or investment outcomes can also be usefully treated as risky situations. The major allure behind conceptualizing or even reducing a complex decision to a risky one is that risk is mathematically tractable. It can be quantitatively characterized by resorting to probability theory, and usefully analyzed by leveraging either frequentist or Bayesian statistics. Whatever we can measure, we can manage and thus—risk management is born.
- **Uncertainty**—this pertains to more complex situations where the outcomes are known or can be defined but their associated probability remains elusive. For instance, the stock market may go up or down, but it is difficult, and very lucrative, to estimate its precise probability of doing so. Indeed, the random walk hypothesis maintains that this is impossible.[17] Decisions under uncertainty are particularly difficult to analyze as unknown probabilities render understanding total impact and tradeoffs all but impossible. This is why such situations are very often reduced to situations of risk by leveraging assumed or approximate probabilities. This allows the theory and practice of risk management to be extended to this domain, as well.
- **Ambiguity**—the final domain of ambiguity pertains to situations where both the probability and the outcomes are unknown and impossible to define. Such situations include rare, and possibly hugely impactful events that are practically immune to prediction. Such situations may include the fallout of a devastating economic collapse or the consequences of climate change or rapid increase in inequality. Such situations of ambiguity closely parallel what Taleb terms "black

---

[13] Again, see Bullen [12].

[14] The concept was pioneered in by Herbert Simon, and its history is nicely reviewed in Klaes and Sent [51].

[15] These are documented by the vibrant field of behavioural economics and aptly summarized by one of its major proponents in Kahnemann [46].

[16] See Taleb [72].

[17] For an overview of how random variables and the idea of random stock movements created the canon of financial economics, please refer to Jovanovic [44].

swans".[18] This is by far the most challenging domain. On the one hand, it is mathematically intractable. On the other, the types of outcomes that lie within are with huge potential impact on both economic and social development. It is thus an important task for risk management to intimately understand ambiguous problems and attempt to define and quantify them better.

Based on this approach the research and practitioner communities have focused their efforts on defining risks and eliciting their change of occurring (probability, p) and their outcomes (impact, x). Given those variables, one can have a point estimate of the expected effect x of a given risk:

$$E[x] = \sum_{i=1}^{n} x_i p_i \tag{1.1}$$

If for example a lucky subject is offered by a researcher to be given a euro each time a coin is tossed and lands tails, then the expected value of this risky bet would be 0.5 €, and thus in ten tosses, the subject would expect to have around 5 €. Of course, even this random event may develop streaks of identical tosses due to the small sample size, and thus the subject could end up with significantly more or less than that.

This early Knightian definition has been hugely successful in that it put the building block of the modern risk management. Sometimes the understanding of risks in economics and business follows the more colloquial one—that of uncertain negative events. This is what is more appropriately called a *pure risk*, whereby all outcomes entail a potential loss. The growing consensus about economic risk is that may include both upsides (positive uncertain outcomes) and downsides (negative ones).[19] Some researchers[20] have taken upon this and divided risks rather into three groups—pure risks that denote exclusively negative events, opportunities which are strictly positive, and uncertainties that may have both positive and negative repercussions. Armed with this understanding, many researchers set out to create a specific and useful typology of risks that can aid its further understanding.

1.2.3 Types of Risks

There are many potentially applicable typologies of risk that seek to delineate the different types of risk and thus—outline the specific domains and situations in which it occurs. Those taxonomies range from very compact[21] that contain a limited set of top-level groups to very detailed ones that explain and classify risk in sometimes

[18] Taleb [72].

[19] Amankwah-Amoah and Wang [3].

[20] Hopkin [42].

[21] E.g. Dionne [30].

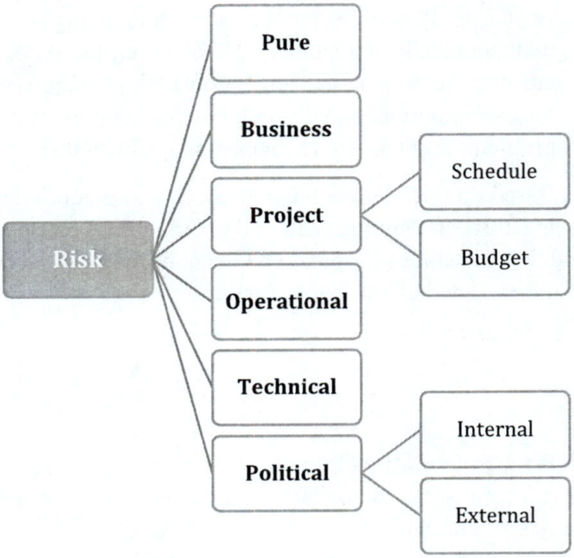

Fig. 1.1 Different types of risks. *Source* Frame [35]

excruciating detail.[22] For the benefit of brevity, we will survey two of the prime examples—the classifications by Frame and Crouhy et al.

Risk Classification by Frame[23]

In his 2003 book J. Davidson Frame proposed one of the most parsimonious typologies of risk that may be of use to economists and managers. It includes merely six large types and purports to cover all relevant aspects for a business (see Fig. 1.1). Those types are as follows:

- **Pure risk**—pertains to exclusively negative events that results in damage or injury. It is also referred to as insurable risk since commercial insurance can be obtained to mitigate it.
- **Business risk**—it is related to the possibility to obtain profit or incur a loss within normal business operations. The magnitude of the risk is directly proportional to the magnitude of the probable outcome—i.e. the greater the amount of risk, the higher its reward must be to compensate the decision-maker for taking it. The typical example of this relationship is the well-known association between risk and expected return in the financial markets. Frame considers business risk as almost identical to entrepreneurial risk which is a somewhat narrower definition than the commonly accepted one.
- **Project risk**—this group encompasses all the risk that stem from the scoping and delivery of projects within the organization. This has mostly to do with scheduling and budgeting risks, whereby projects are significantly delayed or need

[22] E.g. Balabanov [8].

[23] Frame [35].

unexpected additional resources. Common and well-understood projects tend to entail lower level of risk, whereas unique and complex ones, and especially those involving process or product innovations, are generally plagued by higher levels of risk.

- **Operational risk**—those are the risks engendered by the regular operations of a given organization, or, more precisely, that stem from the execution of its business processes. The realization of risk endangers regular activity or support tasks and can be caused by either human error or some mechanical or equipment failure. It is with the growing complexity of operations that such risks increase rapidly.
- **Technological risk**—this grouping includes the risks that are brought about by the introduction of innovations into the organization. Due to the uncertain nature and the novelty of such initiatives it is often difficult to estimate the required timelines, the budget and the overall impact on the organization. All those increase the level of risk.
- **Political risk**—this specific type of risk is due to the influence of political factors (be they internal or external to the organization) on economic decision-making. Internal political risk has to do with power struggles between different individual, resource allocation issues, conflicts of authority and other similar issues. External factors include legal risks as well as the effects of the political process on the economy. Political risk is generally considered lower in developed economies and much higher in developing ones. On the other hand, the returns for taking it on may be significant.

This typology is useful for practitioners and attempts to give a topline idea of the risks that a business decision-maker may confront. On the one hand this gives a parsimonious representation of risk but on the other important groups of risk such as strategic or reputation and only partly and half-heartedly covered under the heading of Business risk. While this may be enough for practical applications, researchers have felt the need to have a further expanded and much more precise typology to guide them in understanding risk.

Risk Classification by Crouhy et al.[24]

A larger and more sophisticated risk typology with a specific focus on economic and business risks is compiled by Crouhy et al. They focus on eight large groups of risks, and further subdivide some of them into smaller sets to fit both practice and regulatory definitions. The typology is briefly presented in Fig. 1.2 and contains the following types of risk:

- **Market risk**—this pertains to the possibility that market dynamics will create favorable or unfavorable conditions with impact on the individual or the organization. An example of that would be how financial market dynamics impact the price of a given asset. This large grouping appropriately contains a number of smaller ones such as the price risk, the interest rate risk, and the forex risk.

[24] Crouhy et al. [25].

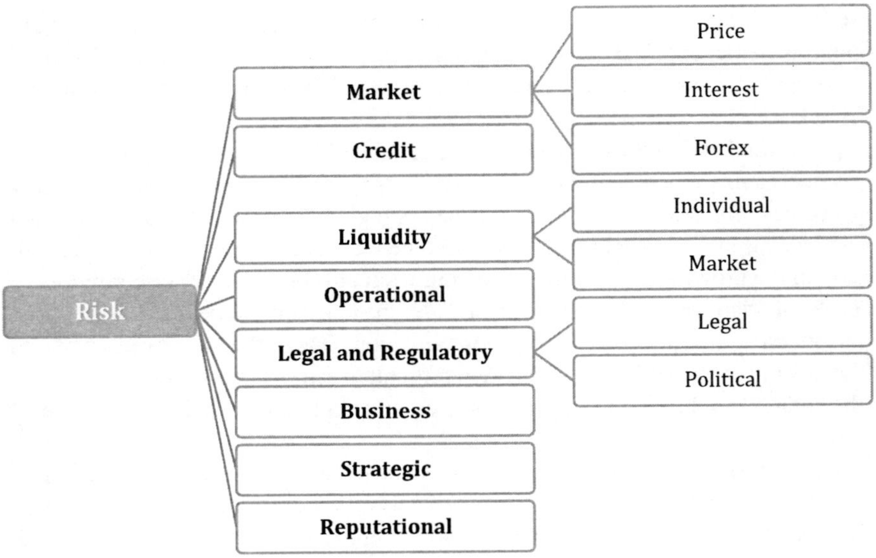

Fig. 1.2 Different types of organizational risks. *Source* Crouhy et al. [25]

Price risk reflects the risk of the change of an asset price on all types of assets—equity, debt, or commodities. Unexpected movements in the level of interest rates engender interest rate risk. Finally, sharp changes in the ratio at which different currencies are traded against each other is what brings about foreign exchange (or forex) risk. Market risks, together with credit risks, have traditionally been in the focus of risk management. Financial sector regulation is partly responsible for this as banks and investment funds are required to measure and report market risk. Additionally, in a world of small data oftentimes data availability defined what can be regulated or researched.

- **Credit risk**—this is another classical type of risk that defined the early development of risk science as understood by economists. Credit risk arises because a given economic agent (e.g. a partner or deal counterparty) is no longer able to service certain debt. This is sometimes reflected in an objective change of this agent's credit rating. The most dramatic credit risk is sovereign risk as governments are unable to service treasury bonds. However, the more mundane type of credit risk—that of individual debtors (people and companies) defaulting on their loans can have equally dramatic effects. It was the latter that brought about the global financial crisis that started in 2007–2008 as large number of mortgage holders could not service their payments anymore. This type of risk is well-studied and well-understood and also the subject of numerous regulations.
- **Liquidity risk**—this is the possibility that the organization or individual will not have sufficient financial resources to carry out planned activities or service an incurred debt. There are two principal types of liquidity risks—individual and market liquidity. In case of individual lack of liquidity, the actor does not have

sufficient financial flow to carry on as intended. This may be due to delays in payment or overall business problems. The second type of liquidity risk occurs when the market itself is not liquid and thus the agent cannot liquidate (turn into money) a given asset, which negatively impacts activities. The practice of cash flow management is particularly sensitive to liquidity risks and aims to minimize them by both real and financial operations. It is worth noting that liquidity risks may be highly correlated with and even cause the realization of other risk, particularly operational and reputational ones.

- **Operational risk**—this is one of the most rapidly growing types of risks modern organizations face. It is connected to possible problems arising within standard operations and generally pertains to issues with business processes, controls, management systems, human error, technology issues, or fraud. Some authors[25] additionally add possibly deleterious external events to this group, as well. The growing complexity of modern economies significantly expand the exposure to operational risk events, and this previously underappreciated group is now garnering significant attention in the research and practitioner communities. Of particular note is that the ongoing process of digital transformation has brought a whole new set of feasible cyber risks, ranging from system failure through information leaks all the way into losing control of key information assets. The growing scope of digital operations are rapidly turning information security risks into one of the most important types of operational risk.[26]
- **Legal and regulatory risk**—those risks stem from unexpected changes in the laws or in the political environment in a given country. Legal changes may mean a lot to a company all the way from increased compliance costs to inability to operate further. In a later article Crouhy et al.[27] also explicitly include political risks such as action to nationalize property or treat given entities in a different way—positively (e.g. through subsidies) or negatively (e.g. through stricter licensing). It is generally considered that political risk is higher in emerging markets with less well-established institutions. On the other hand, legal risk may be considerable in developed jurisdictions as swaths of business are increasingly regulated. This is what happened to the financial sector in Europe and USA after the global crisis, as well as to data processors and biotech companies as the scope of their operations grew.
- **Business risk**—these are classical economic risks that one may find even in very early theories of the firm and consist of the traditional difficulties that arise in the course of normal business operations and market positioning. Those include risk stemming for the inability to precisely forecast market demand, to form precise (equilibrium) pricing, or to accurately plan expenditures. Business risks tend to be on the tactical level of operations and are at the interface between supply and demand. This distinguishes them starkly from operational risks that tend to reflect the company's internal workings. Business risk is somewhat close to the idea of

[25] Chernobai et al. [14].

[26] Egan et al. [31].

[27] See Crouhy et al. [25].

entrepreneurial risk but tends to have a somewhat wider scope as it is focused not merely on new companies or products in search or product-market fit but also on more established ones.

- **Strategic risk**—this is the risk that is inherent in making a large-scale business decision, assuming a significant investment or undertaking any other strategic action with potentially significant impact on the well-being or financials of the actor. Strategic planning, management and positioning activities usually give birth to such risks as they are connected to product innovation, new markets penetration, and strategic changes. Due to their possibly immense impact strategic risks must be a major focus for both organizations and individuals. However, due to a lack of appropriate data these have been under-researched and are usually approached with a set of non-rigorous and even semi-informal methods.[28] A large task for the science of risk is to define, measure and manage strategic risk in a scientifically sound and mathematically tractable way.
- **Reputational risk**—this risk stems from the fact that an actor may be perceived by the public or some relevant counterparty as involved in unacceptable, illegal, unethical, or fraudulent activities. Worsened reputation may lead to a deterioration in the market or competitive positioning of a given organization or individual and thus lead to a weakening of their financial standing. Furthermore, reputational problems may also spell a host of other related issues. For example, as organizations come under public scrutiny and become suspect, regulators may initiate investigations or impose fines, which is what happened to a host of social network companies as the media and public became suspicious of their handling of personal data. Furthermore, reputational damage may spur political or legislative action, or convince counterparties to avoid this actor. Reputational risks are considered "soft" risks as they are notoriously difficult to quantify, and thus their rigorous management remains a standing challenge.

While such typologies are a useful foray for the student of risk, they have a number of inherent limitations. First, a large group of different typologies and taxonomies are in existence, each with its slightly different take on risk, and each on a specific level of detail and granularity. Those are not necessarily comparable and sometimes identical terms have a different meaning or connotation. It is thus of crucial importance for risk science to set on standardized terms. Due to the different communities of research and practice, verbal standardization may be an intractable problem. This is why a more pronounced reliance on mathematical formalization is needed. Second, there is a large and obvious focus on risks that are easier to measure and thus manage. In this respect market and credit risks are much better researched and understood rather than potentially more important risks such as strategic or political risks.[29] Current ability to tackle a problem is obviously a poor measure of its importance. Thus, a second pivotal task for the science of risk is to focus on risks that are difficult to analyze and provide actionable insight into them.

[28] Again, see Hubbard [41].

[29] Ibid.

Third, excessive focus on typologies may leave one under the impression that risks are neatly separated into silos. This is decidedly not the case—different types of risks are highly correlated and one single issue or risk driver may give rise to a number of risk events that are strictly classified under different headings. For example, as Germany's aspiring fintech hopeful Wirecard was exposed in 2020 for fraudulently misreporting a hefty sum of almost 2 Billion euro that went missing, this gave rise to the realization of a number of risks: liquidity risk as counterparties refused to provide loans or pay, operational risk due to internal fraud, and reputational risk as clients started boycotting the company amid regulatory investigation. All those risks are thus highly interconnected and must be managed in an integrated and holistic way. Fourth, it may be useful to focus risk management methodologies to specific industries, thus creating a tailor-made solution that leverages unified horizontal instruments for individual industries and even problems. Some research is under way towards this, and it is already providing interesting results and promising insights.[30] The key task here is striking the appropriate balance between a unified set of mathematically rigorous methods and approaches, and the need for customization on a case-by-case basis.

1.2.4 The Risk Curve

The standard definition of risk focuses on the likelihood of a risky event and its impact, thus attempting to define risks as functions of both probability of occurring as well as attendant consequences. Kaplan and Garrick[31] propose a somewhat expanded scope of the risk definition by considering it as a triplet of factors—a given scenario (denoted s_i) that occurs in real life, the probability of that scenario, p_i, and its impact, x_i. This triplet is then of the following form:

$$\langle s_i, p_i, x_i \rangle \tag{1.2}$$

Risk is thus the set of all possible triplets that exhaust the scenarios which might occur. Thus, we can define risk as:

$$R = \{\langle s_i, p_i, x_i \rangle\}, i = 1, 2, 3, \ldots, N \tag{1.3}$$

A natural critique to this definition harks back to the questions of bounded rationality and limited information that may prevent even the most assiduous analyst from defining every single possible scenario. A proposed option here is to include a catch-all option s_N labeled as "Other" that takes on all residual probability ($p_N = P - \sum_i^{N-1} p_i$). The impact of this miscellaneous category remains much harder to quantify. At any rate, defining every possible outcome is a clear challenge

[30] Molak [57], McManus [55], Ezell et al. [33], Chin et al. [20], Lim et al. [54], Olson and Wu [63].

[31] Kaplan and Garrick [48].

in many complex real-life situations. This definition, on the other hand, has the clear advantage to allow one to visually represent risk over a possibly large group of scenarios. Let us assume a situation of N possible scenarios, each with a well defined triplet of features as per Eq. (1.2), and order all outcomes in ascending order of severity such that:

$$x_1 \leq x_2 \leq x_3 \leq \cdots \leq x_N \quad (1.4)$$

We can calculate the cumulative probability P_i at each scenario that an outcome of at least this magnitude will occur by the following:

$$P_i = p_i + P_{i-1} \quad (1.5)$$

This calculation is summarized in Table 1.1. Effectively the fourth column in row i of the table gives the probability that an event at least as bad as scenario s_i will occur. The astute reader will immediately notice the parallel with the logic behind a distribution function of a random variable. This is by no means an accident. The Kaplan and Garrick definition allows the analyst to think quantitatively of risk, and understand the total expected impact of the complete risk exposure. Table 1.1 is presented visually in Fig. 1.3, that displays a complete risk curve.

The risk curve characterizes a given risk that the individual or the organization faces. It is composed by a number of scenarios, and each of them—with attendant loss (or gain). The major insight from this visual is that there is a large probability of at least some combination of risks occurring—if there are a lot of risks some of them will eventually materialize. It is important to have a complete understanding of the total risk exposure as it may be too easy to dismiss many unlikely risks, while the objective probability of at least some damage stemming from them may be extremely high. Another case in point is that generally risks with high or even catastrophic impact have low probability, but their expected impact may still be very significant. Finally, the risk curve may be used to conceptualize the safeguards needed given the agent's risk appetite. For example, if the organization or individual finds it unpalatable to be exposed to cumulative probability of 80% to take a damage

Table 1.1 Scenario list with cumulative probability

Scenario	Likelihood (probability)	Consequence (impact)	Cumulative probability
s_1	p_1	x_1	$P_1 = p_1 + P_2$
s_2	p_2	x_2	$P_2 = p_2 + P_3$
s_3	p_3	x_3	$P_3 = p_3 + P_4$
…	…	…	…
s_{N-1}	p_{N-1}	x_{N-1}	$P_{N-1} = p_{N-1} + P_N$
s_N	p_N	x_N	$P_N = p_N$

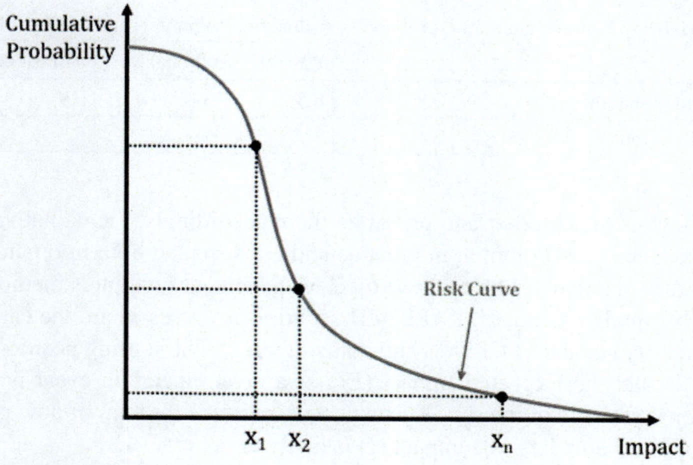

Fig. 1.3 Risk curve. *Source* Kaplan and Garrick [48]

of at least 1 Million Euro, then measures may need to be taken to change the curve parameters.

Despite the usefulness of the risk curve one needs to be aware of its limitations. First, it still offers a somewhat simplified view of risk as scenario outcomes in different domains are lumped together. It may make more sense to have a family of risk curves that reflect different business lines or different types of risks (operational, strategic, business, etc.). Second, the impacts are supposed to be on a single comparable scale so that everything can be evaluated using a single metric. An example of this may be financial costs or benefits but there may be other non-financial outcomes that are more difficult to quantity (e.g. human life, environmental degradation, externalities). The analyst must not shy away from including those difficult-to-measure outcomes but remain vigilant about their exact assessment.[32] Finally, the risk curve reflects a much broader notion of risk than the expected impact. In fact, the expected impact is merely the mid-point or the average of the risk curve. However, a point estimate is much easier to understand, operationalize and communicate and thus using the risk curve may need additional efforts to both construct and apply to an actual risk problem.

1.3 Qualitative Evaluation of Risk

A first pass at estimating risk is to gain an idea of its broad dimensions it qualitative terms, allowing experts to gain initial understanding and obtain a rough idea of how

[32] Still, there is growing interest and understanding on how even intangible values can be easily measured and quantified for the needs of a rigorous analysis. The reader is referred to Hubbard, [40] for more detail.

Table 1.2 Mapping between qualitative probability evaluation, magnitude, and scaling (PMI 2017)

Likelihood	Rare	Unlikely	Possible	Probable	Almost certain
Approximate probability	0.1	0.3	0.5	0.7	0.9
Numeric value (rank)	1	2	3	4	5

risks stack against each other and prioritize them accordingly.[33] Qualitative evaluation methods are very common in practice and are accepted both in organizations and in industry standards such as the Project Management Institute's methodology, the ISO-9001 quality standard, ISO-27001 security risk assessment, the cardholder industry security standard PCI-DSS and many others.[34] The starting position of risk evaluation is that the Expected Impact (*EI*) of a given uncertain event is the key organizational variable of interest. The expected impact is the sum of how probable the risk is (p) and how large its impact (I) is likely to be, or:

$$EI = pI \tag{1.6}$$

Essentially, Eq. (1.6) is a simplified version of Eq. (1.1). The overall risk qualitative risk evaluation exercise proceeds along the following lines:

1. A number of pertinent risks for the process, activity, individual, team, or organization is defined. Large recourse is taken to using subject matter experts and their evaluations, elicited in structured brainstorming, interview, or focus group sessions. Additional care is taken to review existing documents, workflows and schematics, lessons learnt, existing models, and to question different assumptions.
2. The list of identified risks is then entered into a structured form—e.g. in a Risk Log to allow for processing the further evaluation. The probability and the impact of those risks is estimated in qualitative terms. The most common approach by far is to leverage Likert scales to this end. For example, an event's probability may be rated as Rare, Unlikely, Possible, Probable or Almost Certain. Those are then assigned a numeric value (e.g. from 1 to 5) used for further calculations. To ensure consistency among experts, it is usually useful to impose definitions on what approximate magnitudes the qualitative labels refer to. For example, it may be made known that assigning a label Unlikely to an event implies an expert estimate of its probability around 30% (see Table 1.2). In a similar way, the impact of the risk event is also evaluated. The mapping of qualitative impact labels (or Likert scale values) to quantitative values should also play a key part in ensuring consistent estimates.
3. Once probability and impact are evaluated, the analyst obtains the aggregate indicator Expected Impact. This is often done by multiplying the numeric ranks mapped to the qualitative labels. Risks are now prioritized given their Expected

[33] See del Cano and de la Cruz [28], Altenbach [2], Hopkin [42].

[34] Pritchard [65].

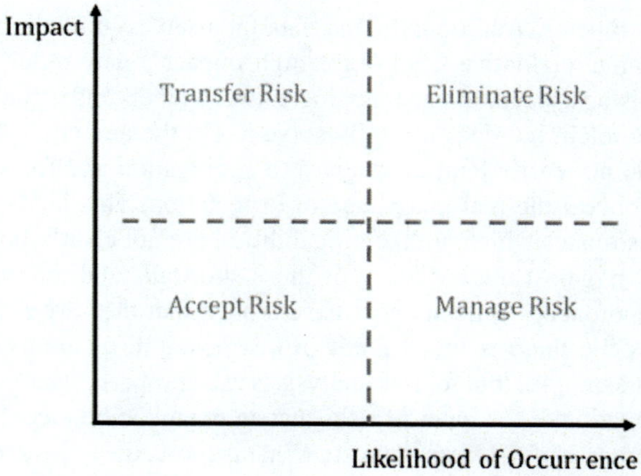

Fig. 1.4 Risk matrix with generic management strategies, based on Ni et al. [59]

Impact. The risk matrix is an oft-used tool to this end. It has the probability estimate on the one axis, and the impact on the other. Risks with highest expected impact fall in the same region of the matrix and are considered the most pertinent ones that need further attention.

4. Based on their position in the risk matrix, analyst and managers are advised to rely on a number of generic risk management strategies[35] (see Fig. 1.4). Risks with a high probability of occurring and a high impact should be eliminated. Those with high impact but low probability can be transferred (e.g. through insurance). Conversely, risks with low impact but high probability may have to be managed internally. And finally, less important risks with a low probability and low impact (thus low expected impact) are to be accepted and merely monitored. Naturally, those are topline strategies and each organization or individual has to put concrete action for each individual risk within the broad framework of those generic strategies.

There are clearly many benefits to using this quick and simple approach to risk evaluation. For once, data collection is relatively easy and requires little preliminary preparation, limited amount of time and no expensive infrastructure. It requires merely the application of a set of well-known and well-understood tools for eliciting and structuring information. Assigning qualitative labels is relatively easy and takes advantage of all the flexibility of using natural language. All this ensures a higher level of cooperation by experts and other stakeholders. The result is a first approximation of risk that is obtained through the use of sparse resources and can be leveraged and communicated widely across the organization.

There are also a number of serious limitations in using qualitative approaches alone. The most common criticism is that qualitative data are mostly subjective and reflect merely the opinion of an expert or a team of experts with nothing to guarantee their validity and reliability. Experts themselves may fall prey to self-serving objectives, thus casting additional doubt upon the quality of the elicited

[35] Ni et al. [59].

estimates. Additionally, the data themselves may not be fully comparable and the same qualitative label (e.g. "high impact") may reflect very different perceptions. Using a standardized mapping between labels and actual values only ameliorates the problem but does not fully solve it. On the analytic side of things, qualitative data do not easily lend themselves to aggregation and formal modeling as they tend to conceal the real magnitude of expectation. This holds particularly true for impact estimates which, unlike probabilities, are not clearly bounded.

Against the backdrop of those shortfalls and the widespread use of qualitative approaches some authors have argued that they are doing more harm than good.[36] While there is much merit in this position, qualitative approaches may still be a meaningful tool for risk analysis is used properly. They are able to give a first approximation of the types of risks that an organization faces and jumpstart a more formal process of risk management. What is more, they sensitize organizations and individuals to the existence of impactful uncertainty in their operations and can facilitate the shift from static deterministic thinking to a more nuanced probabilistic one. The low barriers to using qualitative methods also mean that the inclusion of a much larger group of stakeholders is possible—something that may be desired from both analytic and management perspective. Overall, qualitative approaches are a useful tool for approximations and consensus-building but will likely turn ineffective for precisely analyzing, measuring, and managing sophisticated and large-impact risks.

1.4 A Mathematical Formulation

One possible venue for understanding and modeling risk is by modeling how individual economic agents form preferences and make decisions in situations of uncertainty. Thus, a necessary preliminary is understanding and modeling the agents themselves, and then enriching this with the concept of risk. The idea to formally represent individual preferences in some way is certainly not new, with some authors arguing that it dates back to Aristotelian thinking.[37] Economists have long been infatuated with the idea to rigorously describe preference as this can give much-needed microfoundations for market demand. The venue of choice to do that has traditionally been utility, and its mathematical representation—the utility function.

---

[36] Hubbard [41].

[37] For a discussion, see Kauder [47], particularly Chaps. 1 and 2.

1.4.1 The Utility Function

One of the first to provide a coherent definition of utility was none other than Jeremy Bentham in his 1789 classic *An Introduction to the Principles of Morals and Legislation*,[38] where he stated that utility is "that property in any object, whereby it tends to produce benefit, advantage, pleasure, good or happiness". The concept was later introduced into economics by the efforts of Jevons, Menger and Walras in the 1870s.[39] This concept was significantly refined and later came to the forefront of neoclassical economics only to be later undermined by advances in behavioral and neural economics. Still, the starting premise behind utility remains a useful first pass at understanding how actors make choices. The idea is simple—postulate or elicit preferences and summarize them in a tractable mathematical form. The resulting utility function purports to codify all relevant preferences and desires of the individual that need to be taken into account for a given economic choice. The choice can then be obtained by maximizing this function as the usual assumption is that people seek to achieve maximum satisfaction (utility) out of their choices. This whole process presupposes a specific form of the utility function and certain assumptions on individual preferences.[40]

More specifically, we denote the current or instantaneous utility with u, and assume that it is a function of certain outcomes or results x_i. Those may be outcomes or risky events, but also consumption bundles or even experiences. The utility function is thus of the following form:

$$u = u(x_1, x_2, \ldots, x_n) = u(x_i) \tag{1.7}$$

Hailing from its tradition of modeling consumption choices, it is assumed that the utility function is non-decreasing in its arguments as individuals prefer increased consumption (i.e. more is better), thus leading to a positive first derivative of the utility function with respect to its arguments:

$$\frac{\partial u(x_i)}{\partial x_i} \geq 0 \tag{1.8}$$

On the other hand, the standard assumptions for utility is that people can get satiated from excessive consumption, and every additional unit of the same thing brings less satisfaction than the one before. Mathematically, this can be expressed with a non-positive second derivative of the function with respect to its arguments, or:

---

38 Bentham [11].

39 For a history of utility in its early beginnings up to the middle of the twentieth century refer to Stigler [69].

40 More details in Juster [45].

$$\frac{\partial^2 u(x_i)}{\partial x_i^2} \leq 0 \tag{1.9}$$

The formulation in Eq. (1.9) gives one a first hint of risk preference. It defines a concave function that implies that people will get less satisfaction (utility) as they get more of a given good (e.g. money, status, power) and that they will implicitly tend to prefer certain over uncertain outcomes. In short, classical utility theory supposes individual risk aversion. In all circumstances and all situations, decision makers tend to play it safe. This result, of course, is contradicted by half a century of insights in behavioral economics.[41] We now know that individuals tend to exhibit different types of risk preferences across different decision situations.

The idea of overall mathematical representation of individual preferences is deceptively simple. While one may define an implicit function that is well-behaved, the explicit form of each individual's utility function remains mostly elusive. In fact, the explicit form is non-essential for laying the foundational economic theory. It is much more valuable when concrete modeling, forecasting, and policy recommendations need to be made. One of the popular early explicit formulations is the Cobb–Douglas utility,[42] which has the following form in the case of two goods:

$$u\left(x_{1,}x_2\right) = x_1^{\alpha} x_2^{1-\alpha} \tag{1.10}$$

This function suffers from numerous problems. First, it has no micro-foundations—one cannot ascertain what drive individual preferences and if indeed the drivers will results in a specification of this form. Second, it oversimplifies individual behavior, showing deterministic effects on terminal utility of the different arguments (goods, outcomes, experiences). Third, it fails to account explicitly for risk preferences and cannot accommodate risk taking easily. A number of more advanced and relevant utility functions are reviewed in the sections to follow.

A final comment on utility that merits making is the significance of the difference between instantaneous and total utility. Equations such as (1.7) and (1.10) define the utility derived at one specific moment t by consumption or realization of outcomes. It is thus called instantaneous utility and is hereafter denoted with u. Oftentimes it may be useful to have an idea of the total utility derived over a certain interval (e.g. time). This concept of total utility is denoted by U and can be defined as follows:

$$U = \int_a^b u(x_i)dx \tag{1.11}$$

Both concepts have their use but risk management is often about events at different points in time and therefore total utility over a period is applicable to a wider set of problems in the science of risk.

[41] Kahnemann [46].

[42] Cobb and Douglas [21].

1.4.2 Attitudes to Risk

One of the earliest approaches to formally model risk is to define it as a combination of outcomes and probabilities as per Eq. (1.1) and then model the utility derived from this situation of uncertainty. This can be compared to the mathematical expectation of outcomes and thus see how decision-makers react when having to make the uncertain choice. Risk-neutral individuals care little whether a risk is taken or not but risk conscious ones will factor that into their decision. In short, if the probability of the possible outcomes can be modeled with a well-defined statistical distribution $f(x)$, then the mathematical expectation of the uncertain event is given by the following:

$$E[x] = \int_a^b xf(x)dx \tag{1.12}$$

Empirically, the mathematical expectation is often approximated by the sample mean so that it can be usefully operationalized under realistic situations with readily existing data. Knowing the mathematical expectation one can calculate the utility derived from it as $U(E[x])$ as if this were a certain outcome. On the other hand, agents derive certain utility from the risky outcome, as well. We denote it as $E[U(x)]$ and express it as follows:

$$E[U(x)] = \int_a^b u(x)f(x)dx \tag{1.13}$$

The individual preference between those two quantities in Eqs. (1.12) and (1.13) define three major groups of risk preference and risk behaviors as follows:

- **Risk-neutral behavior**—in this case the expected utility is exactly equal to the utility of the mathematical expectation, or $E[U(x)] = U(E[x])$. Individuals or organizations that exhibit risk neutrality tend to believe that over time the outcomes they experience will tend to converge to their mathematical expectations as they enter many situations of risk. Large financial institutions with billions of transactions can afford to take a risk-neutral approach, and so can Monte Carlo casinos with millions of gambles taking place. Early financial theory supposed risk neutrality is an adequate baseline for rational investors. Advances in behavioral finance, however, showed that this is more of a normative statement than an apt description of reality.[43]
- **Risk-averse behavior**—in this case individuals or organizations derive smaller utility from risky prospects than they would have derived from certain ones of equal magnitude, or $E[U(x)] < U(E[x])$. This is the classical behavior posited in economics and reflected in concave utility curves. People tend to avoid risk and

43 Charness et al. [13].

value certain outcomes higher than uncertain ones, even through their expected results is still the same. This may go even further as economic actors are willing to pay a price to eliminate or decrease uncertainty. Exactly this behavior underlies businesses such as insurance and underinvestment in risky assets despite their superior returns.

- **Risk-seeking behavior**—in this case individual and organizations are actively looking and enjoying risk taking, i.e. they derive greater utility form the uncertain rather than the certain outcome: $E[U(x)] > U(E[x])$. The typical example are thrill-seekers and fans of extreme sports who may be even willing to pay to increase their risks. Particularly bold speculators may also fall in this category.

It is worth noting that those preferences only characterize certain behaviors under a given situation and constraints, and not individuals or organizations as a whole. Thus, it is perfectly expected that one and the same person may be risk averse when it comes to investing all savings into the financial markets and risk seeking for small bets such as playing the lottery. Likewise, a company may be risk neutral when it comes to its regular investments but risk seeking when it budgets its disruptive innovation projects. A large body of research is actually devoted to forecasting what types of situation provoke those different types of behavior. Results are probably best summarized by Dan Ariely's book title that people are predictably irrational.[44] Another useful tool to think about the value (or price) of risk are the so-called certainty equivalents. A certainty equivalent is the certain amount or outcome x_j that results in the same level of terminal utility as the uncertain amount or outcome x_i. The following equality thus holds:

$$U(E[x_i]) = U\left(x_j\right) \tag{1.14}$$

This gives us another perspective on how to measure the price of risk—it is what agents are willing to pay to turn the risky event into its certainty equivalent.

1.4.3 *Incorporating Risk Preferences*

With risk modeling in mind one is tempted to explicitly model risk aversion as part of the utility function, thus rigorously capturing the decision-maker's preferences towards uncertainty. While there are many plausible options, the family of power law utility functions seem to have taken sway in economics, finance, and business applications. Of those, the Constant Relative Risk Aversion (CRRA) specification is decidedly the most popular one.[45] The CRRA utility function belongs to a family of power law distributions that have the following general form:

---

[44] See Ariely and Jones [4].

[45] Wakker [75], Suen [71], Vigna [74].

$$U(x) = \begin{cases} x^r, r < 0 \\ \ln(x), r = 0 \\ -x^r, r > 0 \end{cases} \tag{1.15}$$

In this case r denotes a measure of risk aversion, thus explicitly showing risk preferences and risk tolerance. Additionally, this definition ensures that all functions are strictly increasing with respect to x, as posited in Eq. (1.8). In this family we can see the Constant Absolute Risk Aversion function (CARA) which has an unchanging absolute avoidance to risk for the agent. Thus, it is formulated as follows (with k denoting a parameter):

$$U(x) = -\frac{1}{k}e^{-kx} \tag{1.16}$$

The CRRA specification relaxes thus assumption, and while the level of risk preference does change, the baseline towards which it is estimated indeed does. It follows this specification:

$$U(x) = \frac{x^{1-\theta}}{1-\theta} \tag{1.17}$$

In the CRRA it is the term $1 - \theta$ that governs the level of risk preference of the agent, and θ is the defining parameter. The higher the value of θ, the more risk-averse the individual is. Lower values of θ are associated with risk seeking. Risk aversion occurs at $\theta > 0$, whereas risk seeking is associated with $\theta < 0$. The function is not defined if $\theta = 1$ but its limit at this value approaches $\ln(x)$, which is a useful way to complete its definition.

The question as to the exact value of θ is a clearly empirical one. Chetty[46] states that most economists believe this to be in the range between 1 and 5, Cochrane[47] gives a lower bound of 3, while Azar[48] finds calibrated coefficients to be in range between 4.2 and 5.4. The literature abounds in estimates that are both somewhat lower[49] and somewhat higher[50] than the [1; 5] interval. What is more, there is growing consensus that the coefficients of risk aversion do not remain time invariant but may instead vary significantly.[51]

[46] Chetty [16].

[47] Cochrane [22]

[48] Azar [7].

[49] Guiso and Paiella [38], Denny [27].

[50] Janecek [43].

[51] Conine et al. [23].

1.5 Risk Aversion

Once all relevant preferences are parsimoniously modeled in a utility function that explicitly incorporates risk one can derive both a general representation of a risk aversion metric as well as its explicit form given some utility specification. Such an effort goes beyond assigning a parameter governing risk (such as θ in Eq. (1.17)) and instead tries to construct a deeper model of the drivers of risk preference are. Such an exercise is useful in both scientific and applied settings. In research it provides a framework for the formal study and estimation of risk tolerance. In practice it allows one to characterize agents in terms of their uncertainty avoidance and react appropriately, e.g. when offering them a basket of financial market investments. The Arrow–Pratt measure has been a popular way to characterize risk preference and is thus well-studied, well-understood and amply applied.[52]

1.5.1 The Arrow–Pratt Measure of Risk Aversion

The Arrow–Pratt measure of risk aversion came shortly after von Neumann and Morgenstern's magnum opus on rational-decision making[53] and builds heavily upon these ideas. It also wonderfully coincided with the golden age for the ideas of an economy in general equilibrium fueled by rational choice and optimal markets as embodied in the Arrow-Debreu model.[54] Essentially, two researchers of mathematical inclination, Kenneth Arrow[55] and John Pratt,[56] mounted an analysis of utility functions and derived a rigorous indicator of how risk averse a given individual characterized by this function is. It is instructive to review this derivation and put it in the broader context of understanding risk.

Let us start by assuming an individual whose preferences are summarized by a well-defined utility function $U = U(p_i, x_i)$. The total wealth w_f (or income, or some other variable of interest) consists of a riskless component w_0 and a risky component x, or:

$$w_f = w_0 + x \tag{1.18}$$

The mathematical expectation of the risky component is denoted by μ:

$$\mu = E[x] \tag{1.19}$$

---

[52] Strub and Zhou [70].

[53] Mongenstern and von Neumann [58].

[54] Arrow and Debreu [6].

[55] Arrow [5].

[56] Pratt [64].

Thus, the expected wealth (i.e. the mathematical expectation of total wealth) is given by the following equation:

$$E[w_f] = E[w_0 + x] = w_0 + E[x] \tag{1.20}$$

On the other hand, the expected utility to be derived from this total wealth is represented by the following:

$$U(w_f) = \int_a^b u(w_0 + x) f(x) dx \tag{1.21}$$

At this point we introduce the concept of the certainty equivalent w^*—this is the amount of certain wealth (with a $p = 1$) that will produce the exact same amount of utility $U(w^*)$ as the risky one. The natural question here is what is the amount of this certainty equivalent that will make the individual exactly indifferent between the two options, thus reaching the following equality:

$$U(w^*) = \int_a^b u(w_0 + x) f(x) dx \tag{1.22}$$

The price for the trade between the risky decision and certainty equivalent can be expressed as the difference between the two wealth levels that yield the same terminal utility. Denoting the price as p_a, this can be expressed as follows:

$$p_a = w^* - w_f \tag{1.23}$$

Of course, a risk-neutral decision-maker who cares little about risk can set this price exactly equal to the mathematical expectation of the risky component, or $p_a = E[x]$. However, a risk-averse or risk-seeking individual is bound to ask for a different price which is the risk premium. Denoting the risk premium as η, it is thus:

$$\eta = \mu - p_a \tag{1.24}$$

Positive risk premia reflect risk aversion, while negative ones betray risk-seeking behavior. Substituting the price in Eq. (1.23) into Eq. (1.22), we reach the following expression:

$$U(w_0 + p_a) = \int_a^b u(w_0 + x) f(x) dx \tag{1.25}$$

The two terms $U(w_0 + p_a)$ and $u(w_0 + x)$ can be approximated using a Taylor series expansion. The first one is expanded to the second term, or:

$$U(w_0 + p_a) \approx U(w_0 + \mu) + (p_a - \mu)U'(w_0 + \mu) \tag{1.26}$$

In the expression $u(w_0 + x)$ the potential difference of $(w_0 + x)$ from the point of approximation at $(w_0 + \mu)$ is probably larger than that in Eq. (1.26) which is why it is better to approximate it up to the third term in the Taylor series expansion, or:

$$U(w_0 + x) \approx U(w_0 + \mu) + (x - \mu)U'(w_0 + \mu) + \frac{(x-\mu)^2}{2!}U''(w_0 + \mu) \tag{1.27}$$

Equations (1.26) and (1.27) are substituted in the utility integral (Eq. 1.25). Recall the definition of variance that is also substituted into Eq. (1.25):

$$\int_a^b (x - \mu)^2 f(x)dx = \sigma^2 \tag{1.28}$$

Finally, after some algebraic transformations we reach an expression for p_a and using Eq. (1.24), the resulting expression for risk aversion η is obtained:

$$p_a - \mu = \eta \approx -\frac{\sigma^2}{2}\frac{U''(w_0 + \mu)}{U'(w_0 + \mu)} \tag{1.29}$$

This is the Arrow–Pratt measure of risk aversion. It clearly shows that the larger the variance of outcomes (σ^2), the higher the loss of utility is. Essentially, a risk-averse person is willing to pay more in order to avoid a decision with a wide dispersion of outcomes. The measure also includes specificities of the individual or organizational preferences as captured in their utility function. More specifically, its derivatives reveal the convexity or concavity of the function, thus giving a clear geometric interpretation of risk aversion.

1.5.2 Risk Aversion Under Alternative Utility Specifications

It is sometimes useful, particularly in a modeling context to derive explicit risk aversion measures from a given utility function specification. If the function is not well-defined or well-behaved, this may turn out to be a thorny issue. This is why oftentimes the utility function specification adopted is also considered in view of its mathematical properties and the form of its derivatives. This sometimes comes at the expense of real-life relevance and applicability.

At any rate, the most common contemporary specifications of a utility function used in economics and finance have straightforward and tractable Arrow–Pratt measures.[57] If the modeler has opted for an exponential utility function, such as the Constant Absolute Risk Aversion (CARA) one in Eq. (1.16), then its Arrow–Pratt constant measure of absolute risk aversion (*ARA*) is given by the following:

$$ARA(x) = -\frac{U''(x)}{U'(x)} = k \tag{1.30}$$

A logarithmic utility function has the following form:

$$u(x) = \ln x \tag{1.31}$$

The logarithmic utility function has an even simpler constant relative risk aversion measure (RRA). Its RRA can be expressed as follows:

$$RRA(x) = -\frac{U''(x)}{U'(x)}x = 1 \tag{1.32}$$

Finally, the most commonly used utility specification—the constant relative risk aversion (CRRA) one of Eq. (1.17) has a constant relative risk aversion measure (RRA) defined as follows:

$$RRA(x) = -\frac{U''(x)}{U'(x)}x = \theta \tag{1.33}$$

Despite their popularity, even somewhat sophisticated CRRA utility functions tend to produce unsatisfactory fit when confronted with real world empirical data,[58] which is probably best summarized in Guiso and Paiella's term[59] that when confronting those specifications with data one observes "massive unexplained heterogeneity". Still, the very purpose of modeling is to abstract away from some characteristics of reality which is why it is only expected that utility functions will necessarily fail at capturing every aspect of such a complex process as decision-making under uncertainty by wildly different individuals or organizations. A useful overview of common utility functions and their risk aversion measures is presented in Table 6.1.

The way to approach utility functions may instead proceed in two alternative directions. First, one may use them when the issue under study is well characterized by the utility specification in mind. In this case, even though some aspects of decisions are ignored, those functions may still yield useful and valid conclusions in a constrained domain. A weaker version of this argument is to use those specifications in models where this matters little—quantitatively or qualitatively—but enable the researcher to close the model and converge to a solution. A second route to thinking about utility functions is to consider them as normative prescriptions

---

[57] Vigna [74].

[58] Chiappori et al. [17].

[59] In Guiso and Paiella [38].

rather than as positive descriptions of decisions. Thus, they may provide a useful baseline of what the optimal decision should be even though it is unlikely that every actor makes it. Likewise, the risk preferences and the aversion metrics derived from those utility specifications need to be interpreted carefully and, if possible, calibrated with empirical data.

1.6 Measuring the Risk Premium

Estimating risk in a rigorous and precise manner is both misleadingly simple and unexpectedly difficult. The former rings true as the risk premium can be conceptualized merely as the difference in value (price) between a choice that bears no risk, and an alternative and comparable risky choice. The challenge lies in the many complications in this simple setup—beginning from sourcing and processing data, through factoring in different individual risk preferences, decision context, and time variance of the risk premium. To further illustrate this, we take recourse to measuring the Equity Risk Premium (ERP), using stock market data. The ERP is the difference in price between risky assets traded on the financial markets, and a riskless alternative such as a reference index.[60] Investors may choose to hold a riskless asset such as a stable government bond or take up some risk and buy into equity instead. They naturally need to be compensated for not playing it sure, and the difference in prices shows how much this compensation, or the price of risk, is. Longer time series are usually better for this calculation as random effects tend to cancel each other out. The next sections review different ERPs in Europe and proposes a more sophisticated way of looking at those.

1.6.1 Average Risk Premia in Europe

There is not a single best approach for equity risk premium calculation, but the most common ones tend to fall into one of the following groups[61]:

- **Methods, based on historical data**—this group of approaches tends to leverage existing data on equity returns and bond yields, hypothesizing that the difference between the two over the long run gives a fair estimate of the risk premium. This is predicated on the idea that over time investors conform their expectations to market realities and the averaged actual prices over a sufficiently long period represent a good approximation of the market equilibrium. A possible issue with those approaches is that the analyst tries to measure risk with backward-looking

[60] Siegel [68].

[61] Damodaran [26], Siegel [68].

data, while the relevant risks lie in the future. Should this future fail to faithfully repeat the past, then the estimate may be suspect.

- **Methods based on discounted dividend models (DDMs) or discounted cash flow (DCF) models**—this family of methods largely tries to address the problem of looking at historical data by leveraging forward-looking information. Taking into account discounted future cashflows and current asset prices allows the analyst to calculate an implied risk premium. In this sense, should a given asset be expected to generate certain cash flows *ad infinitum*, then its current price is merely a discounted value of those future cash flows. The rate of discount with proper adjustments is an approximation to the risk premium.
- **Methods based on macroeconomics models**—it is perfectly possible to construct a model of the economy and given the individuals' optimization decisions to calculate what should be the value of the risk premium in equilibrium. As early as the 1980s Mehra and Prescott[62] did just that and found out that the premium was too high given the actual amount of risk undertaken. It was, in fact, so high that it supposed a coefficient of risk aversion of 30 or 40, while the consensus at the time was that it hovers around 1. This came to be known as the "equity premium puzzle" and much literature has ensued trying to solve it. Solutions mostly revolved around the presence of biases in historical data and model improvements. While partially resolved, macroeconomic models are considered a less reliable calculation approach.
- **Survey-based methods**—a final and intuitive way to estimate the ERP is to resort to surveying relevant stakeholders about their expectations. This is rarely the general public, and most often investors, analysts, managers, and academics tend to offer their views in those polls. It seems that survey participants offer surprisingly close number despite their divergent backgrounds, but those numbers diverge significantly from estimates from other methods.[63] Survey results are thus to be treated with extra precaution and triangulated with other sources whenever possible.

Stock market prices and returns are the *par excellence* embodiment of the reward for risk. Selected European stock indices are presented in Fig. 1.5 which amply shows that despite the expectation that risk should be rewarded some markets may grow anemically, remain stagnant, or even decline. Different performances across European markets are a well-established fact (see Table 1.3). Using long-run data for 15 years from 2006 to 2020 shows very wide divergence in both levels of risk and return. The highest average return is observed in non-EU countries such as Ukraine (18%), Turkey and Russia (16%), while in the European Union Norway and Estonia hold the lead with 10% each.

An approximate equity risk premium can be calculated using a reference riskless asset. In Europe an apt choice would be German government bonds which are both widely available to investors and characterized by excellent reputation and credit

---

62 Mehra and Prescott [56].

63 Fernandez et al. [34].

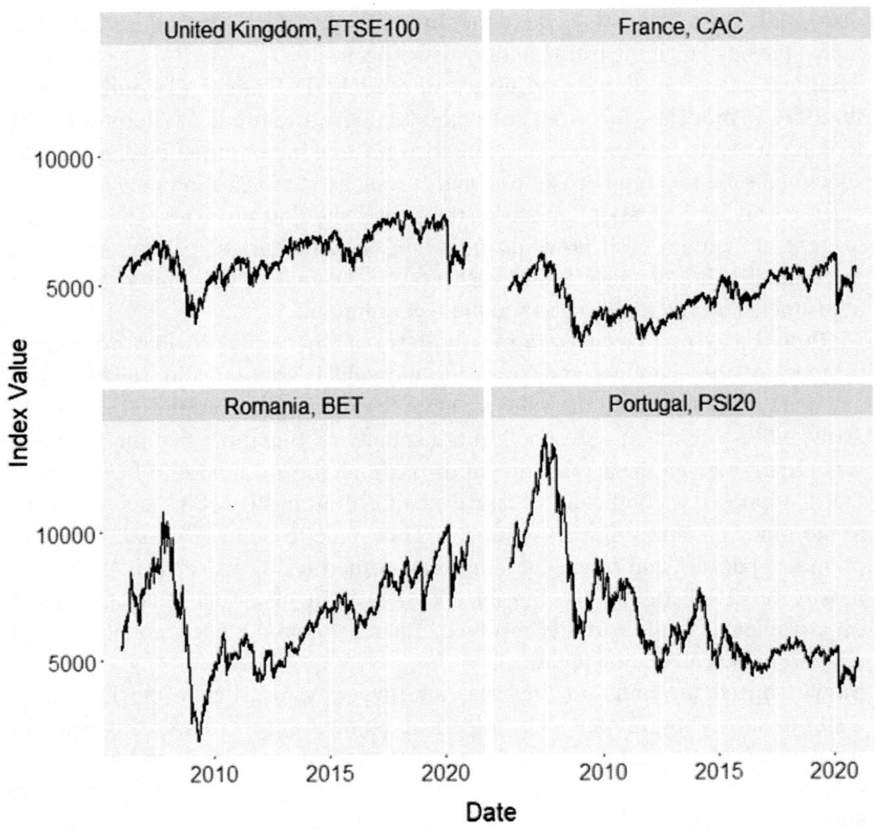

Fig. 1.5 Stock market dynamics in four European countries

rating. Thus, German bonds are a feasible option for a riskless investment across Europe. The reference bond yield for the ERP calculation is taken to be the 10-year bond. The results show a widely divergent financial landscape across Europe. While many countries have a healthy equity risk premium in the range of 5–9% that is common also in other developed countries such as the USA, some results are striking.

A few embattled European economies such as Greece, Portugal, Slovakia, and Spain, have a long-run ERP below zero, showing that stock market investment needs to be avoided there. On the other hand, risky stock markets such the Russian, Turkish and Ukrainian one have a double-digit ERP—16%, 16% and 18%, respectively. This clearly shows the sentiment that risk is higher there and the investor compensation must be accordingly so. The equity risk premium is naturally an approximate indicator of the value or cost of risk, and the methods and assumptions that go into its calculation may lead to widely different results. However, the idea that risk is to be measured by observing actual behavior is indeed a powerful one that is useful and transferable beyond the realm of stock markets.

Table 1.3 Risk, return and equity risk premia across Europe, author calculations

Country	Exchange	Index	Average return (%)	Equity risk premium (%)	Risk (std. dev.) (%)
Belgium	Euronext Brussels	BEL20	2.91	1.42	21.93
Bulgaria	Bulgarian Stock Exchange	SOFIX	3.57	2.07	32.35
Czechia	Prague Stock Exchange	PX	−0.03	−1.52	19.91
Estonia	Estonian Stock Exchange	OMXT	10.22	8.73	31.95
Finland	Helsinki Stock Exchange	HEX	4.65	3.15	21.08
France	Euronext Paris	CAC	2.79	1.29	17.76
Germany	Deustche Boerse	DAX	8.40	6.91	19.61
Greece	Athens Exchange, Athens Composite	ATH	−2.84	−4.33	34.25
Hungary	Budapest Stock Exchange	BUX	8.88	7.39	29.31
Iceland	Iceland All Shares	ICEX	4.00	2.51	29.76
Italy	Borsa Italiano	FMIB	−0.80	−2.29	20.38
Latvia	Latvian Stock Exchange, Riga	OMXR	6.83	5.33	24.94
Lithuania	Vilnius Stock Exchange	OMXV	8.88	7.39	28.07
Netherlands	Amsterdam Stock Exchange	AEX	4.73	3.24	19.77
Norway	Oslo Stock Exchange	OSEAX	9.97	8.47	23.18
Poland	Warsaw Stock Exchange	WIG20	0.51	−0.98	21.49
Portugal	Euronext Lisbon	PSI20	−0.77	−2.27	23.19

(continued)

Table 1.3 (continued)

Country	Exchange	Index	Average return (%)	Equity risk premium (%)	Risk (std. dev.) (%)
Romania	Bucharest Stock Exchange	BET	8.13	6.63	28.97
Russia	Moscow Exchange	MOEX	15.70	14.21	41.03
Slovakia	Slovakia Stock Exchange	SAX	−0.28	−1.78	13.92
Spain	Bolsa de Madrid	IBEX	−0.06	−1.56	19.26
Sweden	Stockholm Stock Exchange	OMXS	6.43	4.94	19.62
Switzerland	Swiss Exchange	SMI	3.55	2.06	15.65
Turkey	Istanbul Stock Exchange	XU100	16.36	13.84	38.54
Ukraine	Ukraine Stock Exchange	UX	18.09	16.59	55.93
United Kingdom	London Stock Exchange, FTSE 100	UKX	1.91	0.42	13.90

1.6.2 Variance Across Countries and Industries

Naturally, risk is an individual-level phenomenon. Each asset or activity has its fundamental level of risk, which is coupled with a specific rate of return that compensates agents for tolerating uncertainty. There are however types and clusters of assets or activities that tend to bear a lower or higher level of risk as compared to other sets. Differing fundamentals tend to be the driving forces behind this phenomenon. As a case in point, the Eurozone is a currency club of mostly highly developed economies with sound fundamentals on both the business and the political front. This results into a lower level of risk as compared to non-Eurozone countries. The flipside of this coin is that the return that investors can hope for in the Eurozone also tends to be lower.

Figure 1.6 displays the density of returns across the Eurozone and non-Eurozone countries in our sample. Countries outside the monetary union have a clear rightward shift of their return distribution, showing that they generate on average greater profitability.

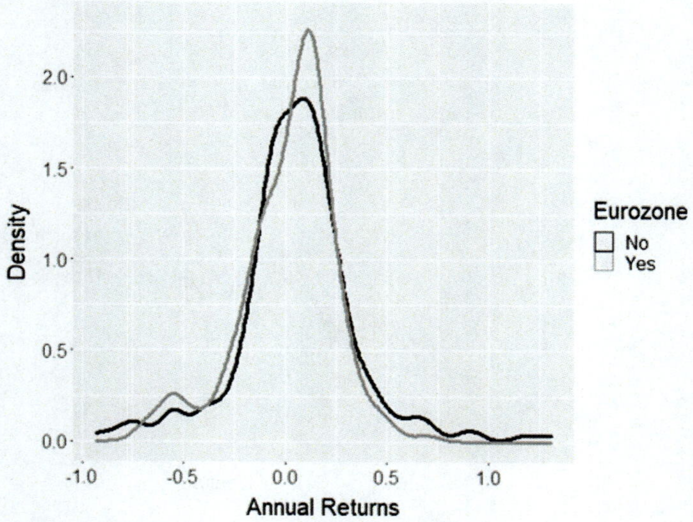

Fig. 1.6 Comparison between returns in and out of the Eurozone

The difference between the two groups of countries also reaches statistical significance at a very high level of $p < 0.0052$ (see Table 1.4), showing that this relation is not a mere visual artefact but a very real tradeoff. In a world with fundamental uncertainty, return is a proxy signal for risk. In the case of financial markets, risk can also be formally calculated using a number of well-known metrics (see Chap. 2) but if actual data in non-existent, the analyst is well-advised to look at the benefits of a given course of action and recognize that higher expected benefits or returns tend to be intricately linked with an increase in the level of risk.

Another level of risk analysis may take place not between countries or regions but within countries. Figure 1.7 shows three German indices—one that aims to capture overall stock market dynamics (DAX), one that focuses on small and medium-sized enterprises (SDAX) and a technology one (TECDAX). All indices are standardized to a 100 as of January 2006. Since the type of companies face different demand and environmental conditions, it is natural that the level of uncertainty in their operations should also differ. SMEs are less resilient to economic shocks due to their size and the scope of their operations and thus riskier. This also leads to a higher growth rate in comparison to the overall DAX index.

Table 1.4 Comparison between average returns in Eurozone and non-Eurozone markets over the period 2006–2020, author calculations

Eurozone mean return	Non-Eurozone mean return	T-statistic	Degrees of freedom	Exact significance
1.84%	5.78%	2.812	389	0.0052

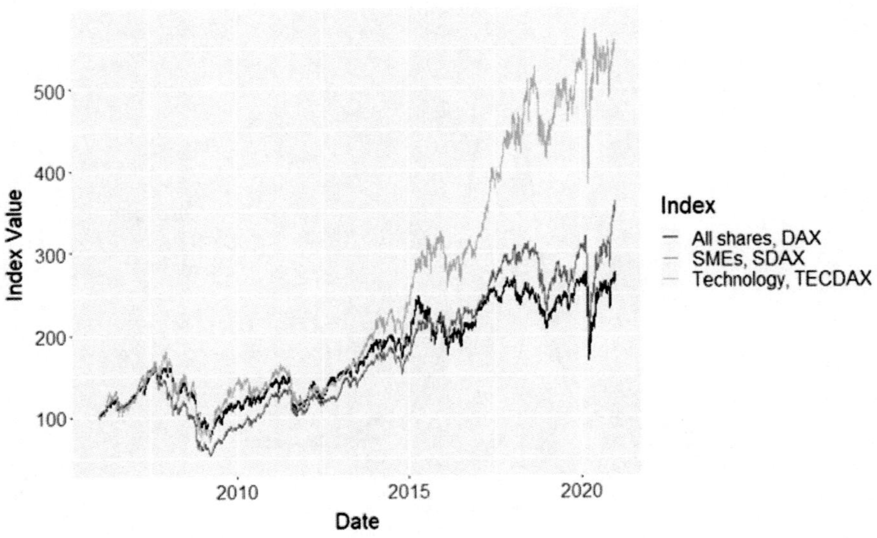

Fig. 1.7 Equity dynamics in Germany: overall index, SMEs and technology

Technology stocks in 2006 tended to be more of a unclear bet. Up to 2015 they perform at par with the overall market as investors fail to appreciate the significant upside inherent in them. As the uncertain process of digital transformation accelerated, and digital businesses were validated, TECDAX started to significantly outperform the market. This aptly demonstrates that there was a significant upside risk inherent in tech stocks. While traditionally risk has been conceived as something deleterious that needs to be steadfastly guarded against, the success story behind technology shows that there are risks on the upside as well that can have a significant effect.

It should be noted that as the global pandemic of 2020 hit the country, the decline in tech stocks was very pronounced and led to reassessment of their fundamental risk-return tradeoff. As a result, by the end of 2020 they have not yet regained pre-crisis levels. In contrast SDAX did so and markedly improved upon that. Some companies within both indices showed divergence from overall movement, again underlining how crucial individual fundamentals are. All this points to the importance of treating risk at the most granular level possible and continuously reassessing the risk exposure as new information becomes available.

1.6.3 *Different Risk Premium Regimes*

While it is widely accepted that the level of risk is a basic characteristic of any activity, asset, or agent, it is also instructive to follow how this level changes over time. The most basic approach is to select different periods and then measure the desired metric over that time. Initially, period selection was done informally by expert evaluation or through a formal statistical test for a structural break such as the Chow test. However, many economic and business processes display a remarkable degree of cyclicality with the boom-and-bust cycle of output being a prime example. The repeating nature of economic states begs the question of whether the analyst can leverage a purely data-driven approach to decide what regime the object of study operates in, and to measure and manage the risk according to those results.

A suitable tool to this end is the Hidden Markov Switching Model approach.[64] The premise behind this model is simple: it assumes that there is a sequence of hidden states (e.g. growth and recession) that one cannot observe but those states generate observable data (e.g. stock returns). This data at hand then allows one to estimate the number and sequence of states, the moments of their statistical distribution such as the mean and standard deviation, and to date them accordingly. A pair of stochastic processes X_n and Y_n ($n \geq 1$) are a hidden Markov process model if X_n are Markov processes, whereby probability P of the current state (S_i) of the process depends only on the previous realizations, and thus the following holds for every n:

$$P(Y_n \in S_i | X_1 = x_1, \ldots, X_n = x_n) = P(Y_n \in S_i | X_n = x_n) \tag{1.34}$$

The Markov model can be further defined with the set of states S, their sequence, and the switching parameters. Most importantly, the model includes a transition probability matrix which shows the analyst not merely the state a process is in, but also the probability of remaining in it or switching to another one. This is highly valuable as it gives an idea of the variable persistence that can be used for further analysis.[65] The Hidden Markov Switching Model can thus give a data-driven determination of what regime the economy (or any system or process) is currently in, and what are the expected levels of risk and return in this specific regime. Additionally, it gives insight on how likely this regime is to continue (i.e. the transition probability). While the use of such models is yet to become a standard tool across all risk management applications, it seems that it is able to consistently outperform alternative forecasting frameworks.[66]

This model is also rather intuitive to apply, requiring only one parameter to estimate—the number of states. While the analyst may choose to use an information criterion such as the Akaike or the Bayesian one, many researchers have opted for assuming a small number of two or three states a priori to ensure model parsimony.[67] Empirical research on optimal number of regimes supports the assumptions that two or three tend to characterize stock markets rather well.[68]

---

[64] Nguyen and Nguyen [62].

[65] For a more detailed description and application to stocks, see Nguyen [60, 61].

[66] Kim et al. [50], Acula and De Guzman [1].

[67] Kole [53], Kim et al. [50].

[68] Dias et al. [29].

Using a standard Hidden Markov Model implementation[69] with the expectation maximization (EM) algorithm, we calculate the parameters of European stock markets[70] under the assumption that they operate in two regimes—bullish and bearish. Results are presented in Table 1.5. The first regime corresponds to periods of market growth with relatively high daily returns—on average up to 0.12% in Isntanbul and 0.09% in the Moscow and Warsaw stock exchanges. The average return is 0.06% with a standard deviation (risk) of 0.93%. In contrast to the bullish first regime, the bearish second is characterized by both a very high volatility, as well as negative expected returns. The average loss across markets reaches 0.12% with the Istanbul, Lisbon and German stock exchanges registering the highest values. The risk in the second regime is also much greater—about 2.5 times higher than in the bullish market. It stands at an average of 2.34% but there are markets in which it goes as high as 3.5% or even almost 4%. It is quite clear that the two states of the market are very different and knowledge of what the current regime is can be very useful for portfolio optimization and risk management.

Remarkably, both regimes are very persistent—the bullish market has an average probability of remaining bullish of over 95%, and the bearish one is just below that. Those results are well in line with much research showing that risk in markets tends to cluster together as high volatility days follow each other. Once the regime changes, a swath of calmer days ensues. The Hidden Markov Model can thus be used as a viable tool for understanding the context of the risk management decision and better calibrating expectations given the current state of affairs. The tool can be both versatile and informative and its use in complex environments is well-warranted.

1.7 Conclusion

As early as the 1980s sociologists such as Ulrich Beck and Anthony Giddens proclaimed that contemporary society is a risk society.[71] In contrast to pre-modern times where risk was a basic feature of the natural environment, modern society is characterized and dominated by man-made risks such as environmental degradation, crime, and nuclear proliferation. The steady pace of civilian and military digital transformation, coupled with advances in biotechnology, makes this observation ever more salient. The increase in socio-economic complexity serves to enlarge the attack surfaces and hence—risk exposures—even further. Yet, the formal analysis and management of risk is yet to be fully embraced as the dominant way to approach political, economic and business problems in a world of uncertainty.

This is not for lack of scientific interest in the subject. The idea of risky decisions can be found even in ancient societies and has been an active field of formal enquiry at

[69] See Sanchez-Espirages et al. [67].

[70] Algorithm reached convergence for 24 of the indices as they display sufficient variance.

[71] See e.g. Beck and Wynner [10] for a more in-depth exploration, as well as Beck [9] and Giddens [36].

Table 1.5 European stocks risk and return under two regimes

Stock exchange	Index	Regime 1 mean (%)	Regime 1 std. deviation (%)	Regime 2 mean (%)	Regime 2 std. deviation (%)
Euronext Lisbon	PSI20	0.06	0.84	−0.21	2.1
Budapest Stock Exchange	BUX	0.06	1.0	−0.07	2.51
Oslo Stock Exchange	OSEAX	0.09	0.90	−0.12	2.4
Istanbul Stock Exchange	XU100	0.12	1.21	−0.24	2.7
Bulgarian Stock Exchange	SOFIX	0.02	0.58	−0.11	2.2
Ukraine Stock Exchange	UX	0.07	1.0	−0.06	3.46
Amsterdam Stock Exchange	AEX	0.07	0.81	−0.14	2.2
Vilnius Stock Exchange	OMXV	0.05	0.42	−0.07	1.9
Euronext Brussels	BEL20	0.07	0.8	−0.14	2.06
Euronext Paris	CAC	0.07	0.86	−0.12	2.2
Moscow Exchange	MOEX	0.09	1.1	−0.09	3.72
Bolsa de Madrid	IBEX	0.05	1.0	−0.11	2.35
Bucharest Stock Exchange	BET	0.06	0.72	−0.06	2.6
London Stock Exchange, FTSE 100	UKX	0.05	0.7	−0.10	1.94
Helsinki Stock Exchange	HEX	0.08	0.86	−0.14	2.1
Estonian Stock Exchange	OMXT	0.04	0.49	−0.02	1.8
Latvian Stock Exchange, Riga All-shares Index	OMXR	0.01	2.38	0.04	0.7
Stockholm Stock Exchange	OMXS	0.08	0.84	−0.09	2.2
Borsa Italiano	FMIB	0.05	1.10	−0.16	2.7
Warsaw Stock Exchange, 20	WIG20	0.09	0.9	−0.13	2.29

(continued)

Table 1.5 (continued)

Stock exchange	Index	Regime 1 mean (%)	Regime 1 std. deviation (%)	Regime 2 mean (%)	Regime 2 std. deviation (%)
Deustche Boerse	DAX	0.07	1.25	−0.20	3.1
Athens Exchange, Athens Composite Index	ATH	0.05	0.81	−0.19	2.6
Prague Stock Exchange	PX	0.07	0.74	−0.17	2.0
Swiss Exchange	SMI	0.04	1.0	−0.08	2.36
Average	**All**	**0.06**	**0.93**	**−0.12**	**2.34**

least as early as the first results in probability theory appeared. The twentieth century, however, marked a high point in our understanding of risk and uncertainty. Beginning from the purely mathematical edifice of utility theory, risk science evolved through the labs of experimental economists and psychologists and is now studied even at the level of individual neurons in the human brain. The basic precept, however, remains the same and allows one to conceptualize risk as an uncertain future event that may bring some relevant outcome. Risks that only entail deleterious consequences are sometimes referred to as pure risks. Following the Knightian tradition, one can also envisage risks that can end in either positive or negative events. Either way, their rigorous analysis and management is key to enhance the former and mitigate the latter.

Formal risk measurement can be based on a simple idea. When people take up a risk, they usually require a certain compensation for their bravery. Thus, the price of a risky bet is an indication of how much this particular risk is worth. The difference between this price and the price of a riskless bet with the same characteristics is the risk premium that allows one to put a numeric (and often monetary) estimate for the value of risk. Risk premia are not time-invariant and respond, sometimes aggressively, to changes in fundamentals or even perceptions. In a world of abundant big data and advanced statistical algorithms it is perfectly possible and even desirable to deploy data-driven approaches to monitoring risk dynamics and analyzing changes. Tools such as Bayesian statistics and models such as the Hidden Markov Model seem particularly promising in this respect. It is through a deeper understanding of the ever more complex risk environment that better management is achieved.

Every activity, agent, or asset has an inherent level of objective risk, which is most likely driven by fundamentals. It is therefore hardly surprising that different country groups, different markets, and different assets show dramatically different levels of risk. One should keep in mind that group belonging can only be considered as a more or less useful first approximation for risk levels. While it is true that government bonds tend to have a lower risk level than company stocks, it is hardly the case that a North Korean bond is a safer bet than a blue-chip US company. This points to the importance of conducting risk analysis at the lowest possible level of granularity. Moving beyond aggregates and understanding the individual building blocks of a given risk exposure may draw the line between success and failure in the risk society. The next chapters proceeds to constructing some basic risk metrics that enable us to better understand and manage the uncertainties and complexities of modern societies.

References

1. Acula, D.D., De Guzman, T.: Application of enhanced hidden Markov model in stock price prediction. J. Model. Simul. Mater. **3**(1), 70–78 (2020)
2. Altenbach, T. J.: A comparison of risk assessment techniques from qualitative to quantitative (No. UCRL-JC-118794; CONF-950740-36). Lawrence Livermore National Lab., CA (United States) (1995)
3. Amankwah-Amoah, J., Wang, X.: Opening editorial: contemporary business risks: an overview and new research agenda (2019)
4. Ariely, D., Jones, S.: Predictably Irrational. Harper Audio, New York, NY (2008)
5. Arrow, K.J.: Aspects of the Theory of Risk-Bearing. Yrjö Jahnssonin Lectures, Helsinki (1965)
6. Arrow, K.J., Debreu, G.: Existence of an equilibrium for a competitive economy. Econom.: J. Econom. Soc. 265–290 (1954)
7. Azar, S.A.: Measuring relative risk aversion. Appl. Financ. Econ. Lett. **2**(5), 341–345 (2006)
8. Balabanov, I.: Risk Management. Moscow: Finance and Statistics Publishing House (1996)
9. Beck, U.: Politics of Risk Society. Environmentalism. Critical Concepts, pp. 256–266. Routledge, London (1998)
10. Beck, U., Wynne, B.: Risk Society: Towards a New Modernity, vol. 17. Sage (1992)
11. Bentham, J.: An Introduction to the Principles of Morals and Legislation (1789), ed. by J.H. Burns and H.L.A. Hart, London (1970)
12. Bullen, E., Fahey, J., Kenway, J.: The knowledge economy and innovation: Certain uncertainty and the risk economy. Discourse: studies in the cultural politics of education, 27(1), 53-68 (2006)
13. Charness, G., Garcia, T., Offerman, T., Villeval, M.C.: Do measures of risk attitude in the laboratory predict behavior under risk in and outside of the laboratory? J. Risk Uncertain. 1–25 (2020)
14. Chernobai, A., Rachev, S. T., Fabozzi, F. J.: Operational risk. Encyclopedia of Financial Models (2012)
15. Chernobai, A., Ozdagli, A. K., Wang, J.: Business complexity and risk management: evidence from operational risk events in US bank holding companies. Available at SSRN 2736509 (2018)
16. Chetty, R.: A new method of estimating risk aversion. Am. Econ. Rev. **96**(5), 1821–1834 (2006)
17. Chiappori, P.A., Salanié, B., Salanié, F., Gandhi, A.: From aggregate betting data to individual risk preferences (2012)
18. Chernobai, A., Ozdagli, A. K., Wang, J.: Business complexity and risk management: evidence from operational risk events in US bank holding companies. Available at SSRN 2736509 (2018)
19. Chernobai, A., Ozdagli, A. K., Wang, J.: Business complexity and risk management: evidence from operational risk events in US bank holding companies. Available at SSRN 2736509 (2018)
20. Chin, A., Kyne, P. M., Walker, T. I., McAuley, R. B.: An integrated risk assessment for climate change: analysing the vulnerability of sharks and rays on Australia's Great Barrier Reef. Glob. Change Biol. **16**(7), 1936-1953 (2010)
21. Cobb, C.W., Douglas, P.H.: A theory of production. Am. Econ. Rev. **18**(1), 139–165 (1928)
22. Cochrane, J.H.: Asset Pricing: Revised Edition. Princeton University Press (2009)

23. Conine, T.E., McDonald, M.B., Tamarkin, M.: Estimation of relative risk aversion across time. Appl. Econ. **49**(21), 2117–2124 (2017)
24. Covello, V.T., Mumpower, J.: Risk Analysis and Risk Management. In: Covello, V.T., Menkes, J., Mumpower, J. (eds) Risk Evaluation and Management. Contemporary Issues in Risk Analysis, vol 1. Springer, Boston, MA (1986)
25. Crouhy, M., Galai, D., Mark, R.: The essentials of risk management (Vol. 1). New York: McGraw-Hill (2006)
26. Damodaran, A.: Equity Risk Premiums: Determinants, Estimation and Implications—The 2020 Edition. SSRN Working Papers, 3550293 (2020)
27. Denny, K.: Upper bounds on risk aversion under mean-variance utility. Available at SSRN 3393383 (2019)
28. Del Cano, A., de la Cruz, M. P.: Integrated methodology for project risk management. J. Constr. Eng. Manag. **128**(6), 473-485 (2002)
29. Dias, J.G., Vermunt, J.K., Ramos, S.: Clustering financial time series: new insights from an extended hidden Markov model. Eur. J. Oper. Res. **243**(3), 852–864 (2015)
30. Dionne, G.: Risk management: History, definition, and critique. Risk Manag. Insur. Rev. **16**(2), 147-166 (2013)
31. Egan, R., Cartagena, S., Mohamed, R., Gosrani, V., Grewal, J., Acharyya, M., Meghen, P.: Cyber operational risk scenarios for insurance companies. Br. Actuar. J. **24** (2019)
32. Evans, A.: Managing Cyber Risk. Routledge (2019)
33. Ezell, B. C., Bennett, S. P., Von Winterfeldt, D., Sokolowski, J., Collins, A. J.: Probabilistic risk analysis and terrorism risk. Risk Anal. Int. J. **30**(4), 575-589 (2010)
34. Fernandez, P., Aguirreamalloa, J., Avendano, L.C.: US Market Risk Premium Used in 2011 by Professors, Analysts and Companies: A Survey with 5,731 Answers. SSRN Working Papers, 1805852 (2011)
35. Frame, J. D.: Managing risk in organizations: A guide for managers. John Wiley & Sons (2003)
36. Giddens, A.: Risk and responsibility. Mod. Law Rev. **62**, 1–10 (1999)
37. Grier, B.: The early history of the theory and management of risk. In: Judgment and Decision Making Group Meeting, Philadelphia, USA (1981)
38. Guiso, L., Paiella, M.: The role of risk-aversion in predicting individual behavior. In: Chiappori, P.-A., Gollier, C. (eds.) Competitive Failures in Insurance Markets: Theory and Policy Implications, pp. 213–250. MIT Press and CESifo (2006)
39. Guiso, L., Paiella, M.: Risk aversion, wealth, and background risk. J. Eur. Econ. Assoc. **6**(6), 1109–1150 (2008)
40. Hubbard, D.W.: How to Measure Anything: Finding the Value of Intangibles in Business. Wiley & Sons(2014)
41. Hubbard, D.W.: The Failure of Risk Management: Why It's Broken and How to Fix It. Wiley (2020)
42. Hopkin, P.: Fundamentals of risk management: understanding, evaluating and implementing effective risk management. Kogan Page Publishers (2018)
43. Janecek, K.: What is a realistic aversion to risk for real-world individual investors. Int. J. Financ. **23**, 444–489 (2004)
44. Jovanovic, F.: The construction of the canonical history of financial economics. Available at SSRN 3294557 (2018)
45. Juster, F.T.: Rethinking utility theory. J. Behav. Econ. **19**(2), 155–179 (1990)
46. Kahneman, D.: Thinking, Fast and Slow. Macmillan (2011)
47. Kauder, E.: History of Marginal Utility Theory. Princeton University Press (2015)
48. Kaplan, S., Garrick, B. J.: On the quantitative definition of risk. Risk Anal. **1**(1), 11-27 (1981)
49. Keynes, J.M.: A Treatise on Probability. Macmillan, London (1921)
50. Kim, E.C., Jeong, H.W., Lee, N.Y.: Global asset allocation strategy using a hidden Markov model. J. Risk Financ. Manag. **12**(4), 168 (2019)
51. Klaes, M., Sent, E.M.: A conceptual history of the emergence of bounded rationality. Hist. Polit. Econ. **37**(1), 27–59 (2005)
52. Knight, F.: Risk, Uncertainty and Profit. Hart Schaffner, and Marx (1921)

53. Kole, E.: Markov Switching Models: An Example for a Stock Market Index. SSRN Working Papers, 3398954 (2019)
54. Lim, S. S., Vos, T., Flaxman, A. D., Danaei, G., Shibuya, K., Adair-Rohani, H., ... & Aryee, M.: A comparative risk assessment of burden of disease and injury attributable to 67 risk factors and risk factor clusters in 21 regions, 1990–2010: a systematic analysis for the Global Burden of Disease Study 2010. The Lancet, **380**(9859), 2224-2260 (2012)
55. McManus, J.: Risk management in software development projects. Routledge (2012)
56. Mehra, R., Prescott, E.C.: The equity premium: a puzzle. J. Monet. Econ. **15**(2), 145–161 (1985)
57. Molak, V.: Fundamentals of risk analysis and risk management. V. Molak (Ed.). Boca Raton, FL, USA: Lewis Publishers (1997)
58. Morgenstern, O., Von Neumann, J.: Theory of Games and Economic Behavior. Princeton University Press (1953)
59. Ni, H., Chen, A., Chen, N.: Some extensions on risk matrix approach. Saf. Sci. **48**(10), 1269-1278 (2010)
60. Nguyen, N.: An analysis and implementation of the hidden Markov model to technology stock prediction. Risks **5**(4), 62 (2017)
61. Nguyen, N.: Hidden Markov model for stock trading. Int. J. Financ. Stud. **6**(2), 36 (2018)
62. Nguyen, N., Nguyen, D.: Hidden Markov model for stock selection. Risks **3**(4), 455–473 (2015)
63. Olson, D. L., Wu, D. D.: Enterprise risk management models. Heidelberg: Springer (2017)
64. Pratt, J.W.: Risk aversion in the small and in the large. Econometrica **32**(1/2), 122–136 (1964)
65. Pritchard, C. L., PMP, P. R.: Risk management: concepts and guidance. Auerbach Publications (2014)
66. Project Management Institute, PMI.: Project Management Body of Knowledge (PMBOK Guide), 6th Edition. US: PMI (2017)
67. Sanchez-Espigares, J.A., Lopez-Moreno, A., Sanchez-Espigares, M.J.A.: Package 'MSwM'. Traffic **15**(1) (2015)
68. Siegel, L.B.: The Equity Risk Premium: A Contextual Literature Review. CFA Institute Research Foundation (2017)
69. Stigler, G.J.: The development of utility theory. I. J. Polit. Econ. **58**(4), 307–327 (1950)
70. Strub, M.S., Zhou, X.Y.: Evolution of the Arrow-Pratt measure of risk-tolerance for predictable forward utility processes. Available at SSRN 3276638 (2018)
71. Suen, R.M.: Bounding the CRRA Utility Functions. University Library of Munich (2009)
72. Taleb, N.N.: The Black Swan: The Impact of the Highly Improbable, vol. 2. Random House (2007)
73. Ulam, S.: Adventures of a Mathematician. Charles Scribner's Sons, New York (1976)
74. Vigna, E.: Mean-variance inefficiency of CRRA and CARA utility functions for portfolio selection in defined contribution pension schemes. Collegio Carlo Alberto Notebook 108, or CeRP wp 89, 9.
75. Wakker, P.P.: Explaining the characteristics of the power (CRRA) utility family. Health Econ. **17**(12), 1329–1344 (2008)

Chapter 2
Standard Risk Metrics

2.1 Introduction

The existence of risk in economic activity is hardly breaking news. What is surprising is its overwhelming role in business and social decisions and, sometimes, the incredible propensity of individuals and organizations to ignore this fact.[1] Against the backdrop of increasing economic sophistication and ever-expanding risk exposure, particularly to high impact risks, it is of paramount importance to understand and rigorously manage the inherent uncertainty in practically every action we take. This chapter begins modeling risk not as an individual-level attribute (as with risk preferences derived from a utility function) but rather as a systemic property that needs to be understood and governed at a systemic level. To this end one can heavily rely on probability theory but also never substitute it for sound reasoning.

A crucial first preliminary observation before this study is that risk exists objectively in the economic, social, environmental and other systems. Every risk metric used is an attempt to operationalize and measure this objective variable, and it thus potentially fraught with numerous measurement problems. This may lead to two distinct pitfalls—overreliance on and skepticism about risk metrics. The former is usually brought about when sophisticated mathematical methods are used that generate metrics with a seemingly very high degree of precision, which in turn mislead decision makers to believe they have accurately measured a risk and can thus decrease buffers and contingencies. In fact, they have most probably obtained an imprecise point estimate that may vary across a wide confidence interval.[2] In short, the risk metric is also risky. This brings one to the other pitfall—the inability

[1] See e.g. Silver [55].

[2] This point is thoroughly made and illustrated with the story of the spectacular fall of probably one of the most quantitatively sophisticated hedge funds—Long Term Capital Management (LTCM), in Rahl [49].

A. Gerunov, *Risk Analysis for the Digital Age*, Studies in Systems, Decision and Control 219, https://doi.org/10.1007/978-3-031-18100-9_2

to reach ultimate precision may lead to outright rejection of any measurement to the detriment of the decision at hand.[3] However, one should remember that the purpose of measurement is to decrease uncertainty to the level needed for a decision, and not to eliminate any uncertainty at all (even if that were possible). So, risk analysis is like any other measurement—extremely useful if done well but still a somewhat imperfect representation of a complex underlying phenomenon.

Second, one needs to note that economic agents comprehend risk within the constraints of the information they dispose of, their cognitive resources, and the decision architecture within which they are placed.[4] This leads to the simple observation that while risk exists objectively in the natural (and digital) world, agents perceive it subjectively, incorporating their values and biases. Most often, the objective and subjective risk fail to coincide. The distinction between the two may be viewed also as a philosophical one as some authors have noted[5]—indeed, all we can see, and measure is necessarily blurred by the lens of subjective instruments, values, and judgments. As a more practical research matter, however, it is useful to retain the distinction between the two to aid their better study. One can thus define objective risk as the empirically existing and (imperfectly) measurable risk, while subjective one is how individuals and organizations see and conceptualize it.

Third, the consensus of what the appropriate metric to risk is may shift unexpectedly, and often dramatically. While all risk metrics may have theoretical shortcomings, at a given point in time some are considered the state-of-the-art and are thus both extensively researched and applied in practice. For a period of time into the 2000s such a metrics was the Value at Risk (VaR) one that shows what a minimum expected loss with a certain probability over a given period. While its limitations were known, it was heralded as a major progress in the field and endorsed (and enforced) by key regulators, including the Basel Committee. Under the stress of the financial crisis of 2007–2008 and the growing weariness of the public, academia, and communities of practice, this endorsement was reversed in favor of the equally controversial Expected Shortfall metric that is supposed to give a better notion of the probability of extreme events.[6] Obviously, objective existing risk did not change as a result of that but the way it is measured, reported, and understood did. The analyst must be cognizant of the possibility of such shifts and instead of relying on a single fashionable metric resort to a wider spectrum of relevant ones. This chapter provides an overview of a number of common ones.

---

[3] McGoun [40].

[4] For an enticing overview of the importance of decision architectures, the reader is referred to Thaler and Sunstein [57].

[5] Hansson [29].

[6] Rowe [51].

2.2 Risk as a Random Variable

The intuitive way to think of risk is as a one realization out of many uncertain possibilities. It is thus only natural to mathematically model risk as a random variable whose current realization (the outcome one is currently looking at) is drawn at random from its respective distribution. Recalling that the statistical distribution of a random variable completely characterizes its probability of realizing certain values, it follows that utilizing it to model risk gives a full picture of its likelihood that can be used for expectations formation and decision analysis. Naturally, this is an oversimplification. Probability distribution functions tend to be continuous curves, implying that there are no discontinuities or jumps in the likelihood of a given risk occurring. In addition, real-life behavior of risky outcomes may have a number of regime changes (inflection points) that are hard to capture in the common distributions. Finally, the distribution function is a model of the behavior of the generating process and may or may not have very good fit to the actual data at some points. This is particularly true at the distribution tails (very high or very low probability events) that usually characterize non-standard edge cases. However, it is oftentimes those edge cases that have a disproportionate impact on the risk exposure.

That said, leveraging a suitable distribution may greatly aid the measurement and analysis of risk, and, within the constraints of its fit, generate useful conclusions. Despite the wide variety of statistical distributions known, the large number of risk management research and application tends to focus on a few families that are both mathematically tractable and posses a number of desired properties. We review four of those distributions (or families of distributions) and see how the their parameters may be tweaked to reach a better fit to data. This is usually the easier part of modeling and can be done using automated algorithms for best fit. The challenge remains in choosing the type of distribution that is appropriate for this specific use case in the first place.

2.2.1 *Normal Distribution*

The undoubted classic in modeling is the Normal (or Gaussian) distribution. It is a particularly good fit for variables that have a well-defined middle point and tend to symmetrically fall around it, with an ever-decreasing likelihood of them falling further away. This would be the case in some physical quantities such as height, weight, intelligence. Essentially, the Normal distribution is well-suited for describing variables that tend to be bounded from above or below as it gives very low probabilities to extreme values. It is defined as follows:

$$f\left(x|\mu,\sigma^2\right) = \frac{1}{\sigma\sqrt{2\pi}}e^{-\frac{(x-\mu)^2}{2\sigma^2}} \tag{2.1}$$

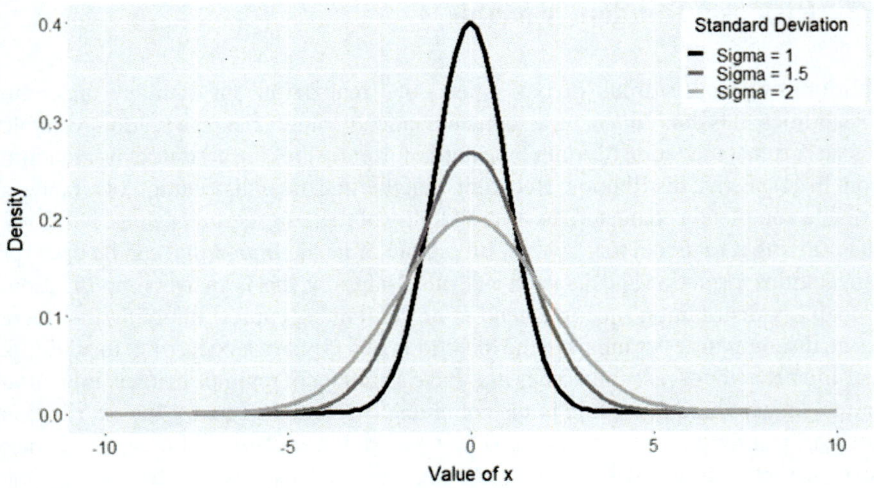

Fig. 2.1 Normal distributions with different standard deviations

Equation (2.1) is plotted graphically in Fig. 2.1.

The key parameter that denotes risk in the normal distributions is the standard deviation (most often denoted as sigma). The larger the standard deviation, the more dispersed the realizations of the random variable around its mean, and thus—the higher the risk. The Gaussian distribution is highly sensitive to variations in its standard deviation—as Fig. 2.1 vividly shows even small increments (from one to two or three) result in significantly different distributions. In the final analysis, even small tweaks of the distribution's standard deviation will lead to large numeric differences in estimates that use it. Irrespective of the precise shape and form, the Normal curve still supposes low probabilities for tail events and is thus an excellent choice for modeling trivial phenomena.

It shines when applied to variables that are clearly bounded from above and below, cluster around a central tendency, and exhibit limited non-linearity. The classic example here is human height—most humans cluster tightly around the average height which varies globally between 1.6 and 1.8 m. There are some people of exceptionally high stature but the highest among them reaches 2.72 m.[7] There is no recorded evidence of statures of three meters. The situation is similar with the lowest height. In short, genetics, nature, and metabolic laws put a strict limit above and below, and lead to an overwhelming clustering in the middle. In economics and business, we are likely to observe similar distributions with well-known standardized processes.

More insidiously, researchers and practitioners use the Gaussian curve to model variables that resemble, are close enough, or are likely to converge to a Normal distribution. Thus, this is the default option. Many an argument has raged whether

[7] For example, there is recorded evidence for Robert Wadlow, known as the giant of Illinois, who reached 2.72 m.

this approximation is good enough for academic and practical purposes, and this has rarely been more involved than in the case for financial market returns. On the one hand, many researchers have been actively using the normal distribution for stock market returns or, at the very least, failing to reject normality.[8] On the other hand, many others have argued that the Gaussian is inadequate and have proposed other alternatives.[9] In particular, the Normal distribution is most questionable when used for modeling rare events such as catastrophic stock market crashes, global pandemics, and violent political and economic turmoil. The Gaussian curve that tends to put extremely low probability on those is sometimes not borne out in practice.

2.2.2 *Family of Fat Tails Distributions*

Sticking to the baseline of normality is appealing, and indeed useful, when dealing with the overwhelming proportion of risks. These are mostly minor incidents, part of day-to-day operations that are situated around the peak of the Gaussian distribution and are well-described by it. Rarely are those uninteresting risks the focus of ground-breaking research or tectonic shifts in the economy, society, businesses, or the environment. It is often the rare little understood risks that lie at the tails of the distribution that capture our imagination and give rise to significant impacts. Those so-called tail risks are sometimes poorly described by the normal distribution as it implies that their probability is extremely low. Often, it is not. Nassim Taleb provocatively put it that seeing a six-sigma (an extremely rare) event is evidence that it is not really a six-sigma event. Wall Street traders are even more blunt, joking that they witness six-sigma events (with a probability of 0.0034%) every other month. In short, the normal distribution may underestimate rare high-impact events.

A large body of research in finance and risk management have taken up those anecdotal evidence to heart and started to look rigorously into it. Numerous articles have found that some aspects of the financial markets are much better described by distributions that give higher probabilities to rare events.[10] Visually, this means that the tails of the distribution are fatter than the normal ones, and thus the name fat tails distributions is born. Two common distributions to consider for our risk modeling is the Cauchy and the Student T distributions.

The Cauchy distribution is a popular one within the natural sciences and is a well-defined stable distribution, largely governed by a scale parameter γ. If we denote the peak of this distribution with x_0, then its probability density function has the following form (Fig. 2.2):

[8] Akter and Nobi [3], Ling [37], Liu et al. [38].

[9] Peiró [44].

[10] Daníelsson et al. [13], Eom et al. [16], Huisman et al. [31], Verhoeven and McAleer [59].

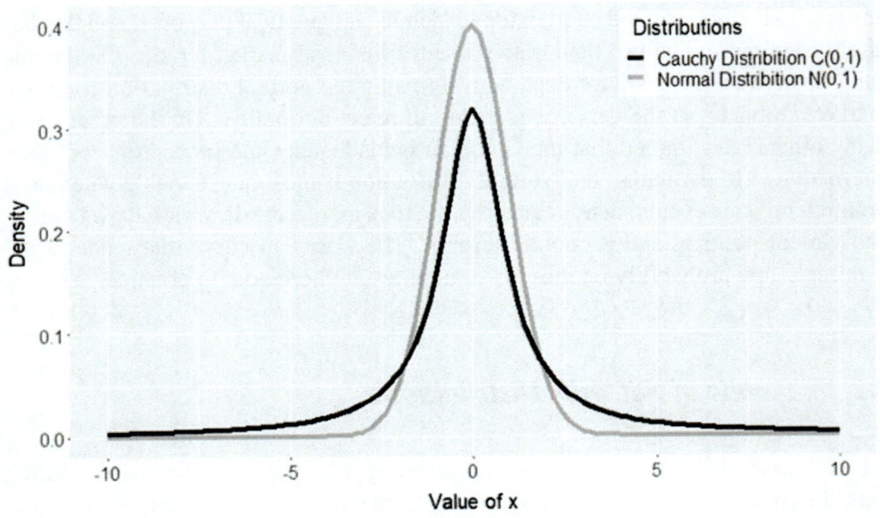

Fig. 2.2 Comparing a normal and Cauchy distribution with similar parameters

$$f(x|x_0, \gamma) = \frac{1}{\pi\gamma}\left[\frac{\gamma^2}{(x - x_0)^2 + \gamma^2}\right] \tag{2.2}$$

The Cauchy distribution provides a notably higher probability of extreme events and can thus be employed to ensure that the chance of a high impact risk occurring is not underestimated. As a direct comparison, we can see the differences between a Cauchy distribution with location 0 and scale of 1 as compared to a Gaussian distribution with a mean of 0 and standard deviation of 1 in Fig. 2.2. Thicker tails for the Cauchy mean than there is less probability for realizations around the mean, but more for outliers. Thus, numeric estimates for rare high impact events will be significantly different.

Another popular fat tails distribution is the Student T. Its shape is governed by the degrees of freedom parameter that we denote with ν. This is dependent on the sample size, and $\nu = n-1$. To define it we also take recourse to the Gamma function, denoted as Γ (where $\Gamma(n) = (n - 1)!$). The probability density function is as follows:

$$f(x) = \frac{\Gamma\left(\frac{\nu+1}{2}\right)}{\sqrt{\nu\pi}\,\Gamma\left(\frac{\nu^2}{2}\right)}\left(1 + \frac{x^2}{\nu}\right)^{-\frac{\nu+1}{2}} \tag{2.3}$$

Visually, the Student T distribution much more resembles the normal one (see Fig. 2.3).

However, its key benefit lies in its versatility. Given different values of its degrees of freedom parameter ν, it converges to other types of distributions. At $\nu = 1$, it is the same as the Cauchy distribution, and at $\nu > 30$ it starts approximating the Normal one. Finally, at intermediate values of ν between 1 and 30, its probability density function

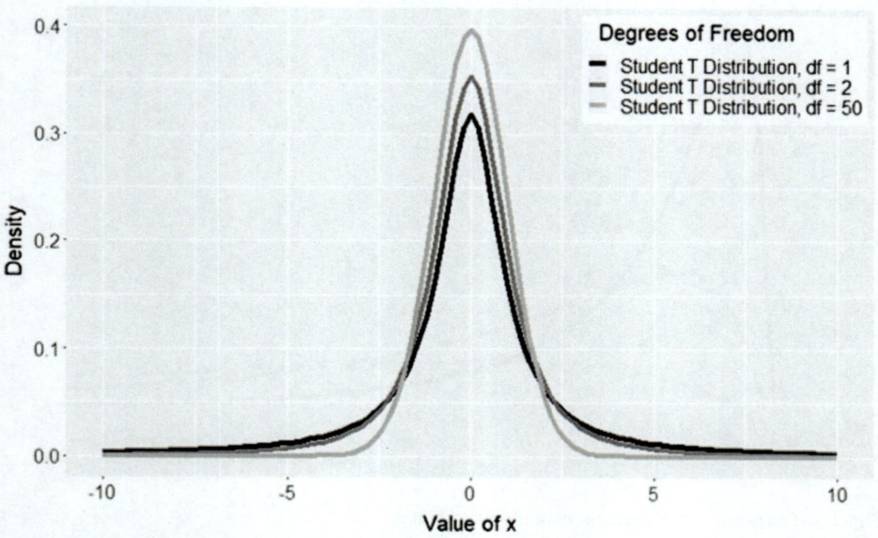

Fig. 2.3 Student T distributions with different degrees of freedom

generates a family of fat tail distributions. The optimum value of the parameter may also be set given a dataset and performing a grid search for the value that best fits data at hand.

2.2.3 Log-Normal Distribution

Empirical observation of economic and business processes has shown some intriguing features many of them share. Given a certain process, like payments, the overwhelming majority of transactions tend to cluster around a well-defined mean and form a somewhat symmetric bell around it, but there are also some rare events (e.g. severe recession) that provide for a very long tail of the distribution. Such events may be catastrophic deviations that do not often occur but have significant impact when they do. Those characteristics can be aptly captured by a log-normal distribution. Indeed, real-life situations in field as diverse as biology, sociology, and economics can well be approximated by the specific shape of the lognormal.[11] This distribution is defined in the following Eq. (2.4):

$$f(x) = \frac{1}{\sigma\sqrt{2\pi}(x-\theta)} e^{-\left(\frac{(\ln(x-\theta)-\mu)^2}{2\sigma^2}\right)} \tag{2.4}$$

---

[11] Gualandi and Toscani [26].

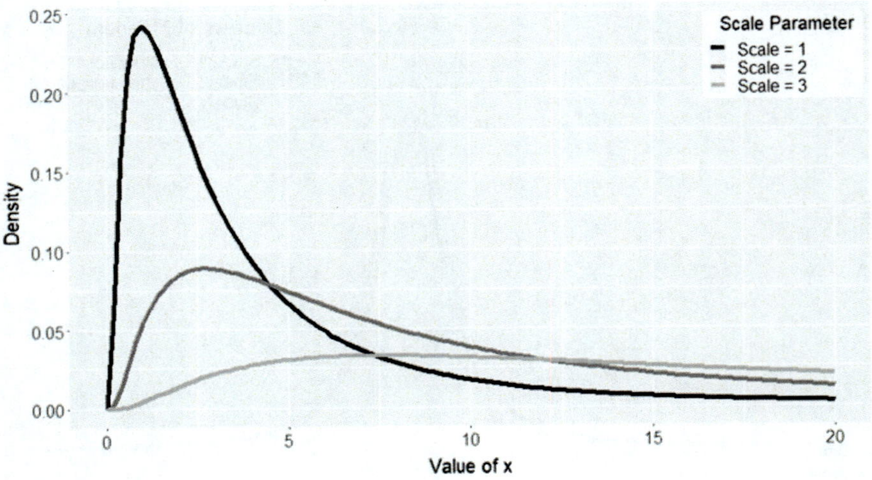

Fig. 2.4 Lognormal distributions with different scales

The distribution is defined by its two parameters—scale, μ and shape, σ, and to a somewhat lesser extent—by its location parameter θ. The lognormal distribution is visualized in Fig. 2.4.

One of the most apt applications of the lognormal distribution is for modeling operational risks. Those are common risks that every modern organization faces and stem from the potential problems associated with people, processes, systems, and external events.[12] The total operational risk is thus composed of many regularly occurring errors and accidents (e.g. in terms of data entry and processing, erroneous transaction approvals, minor fraud, unimportant human error, etc.) that form a high initial peak of the distribution. The nature of those events results in rather limited losses associated with them. However, at points a catastrophic failure occurs such as a large-scale data breach, ransomware attack, or large-scale fraud by executives. Those unlikely calamities form the very long tail of the distribution. The losses are further magnified by the correlation between the different types of risks that occur together.

The Equifax data breach of 2017 is an apt case in point.[13] Initially, the company failed to patch a known security vulnerability (a relatively minor IT error at the time) which three months later led to the exposure and compromise of highly sensitive personal and financial data for about 145 Million customers (catastrophic IT risk). Equifax failed to notify the public and regulators immediately for the breach and top executives use the time before announcement to sell stock in the company, prompting suspicion for fraudulent insider trading—a people-related risk that materialized only because of the IT risk that happened.[14] In addition to Equifax's shattered reputation,

[12] Chernobai et al. [10], Leone et al. [36].

[13] Berghel [5].

[14] Safi [53].

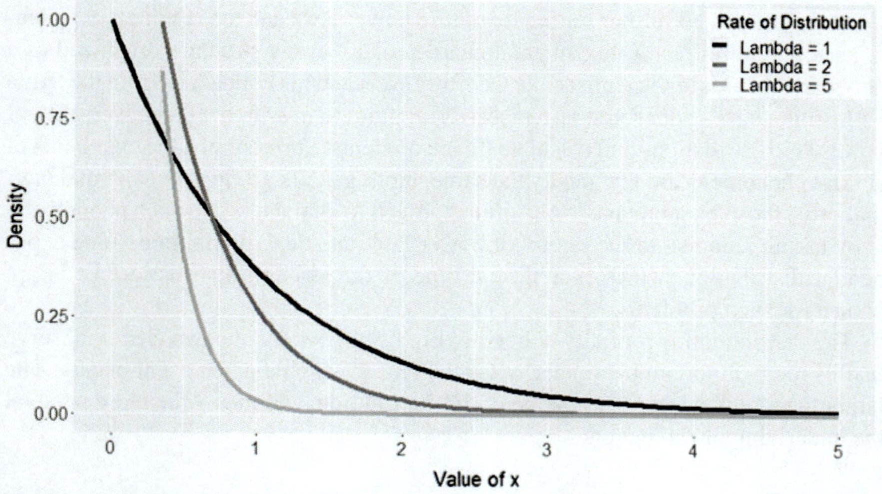

Fig. 2.5 Exponential distributions with different rates

the company had to settle in 2020, paying out 452 Million dollars. While the overwhelming number of risks that Equifax faced up to this point were minor errors and small-scale fraud, the long tail of the distribution had a few unpleasant (and highly-correlated) surprises in store for them. One can thus view the lognormal distribution as a better choice than the normal one for modeling potential rare and negative risky events as it features a very long one-sided tail.

2.2.4 Power Law Distributions

In 1896, an Italian engineer who turned to economics and was at the time teaching at the University of Lausanne was about to make a crucial observation about distributions of real-life phenomena. Vilfredo Pareto has recently succeeded Leon Walras as the Chair of Political Economy at the university and was currently investigating patterns of wealth ownership. He noted that 80% of the total wealth was in the hands of about 20% of the population—the 80/20 principles that was later named after him. In short, the overwhelming majority of people have rather little wealth, whereas a tiny minority are extremely rich—i.e. wealth follows a power law distribution such as the one shown in Fig. 2.5. As it turns out, many phenomena in economics and business are characterized by a similar shape, reflecting a winner-takes-all dynamic.[15] Most notably, power laws are observed in the distribution of city sizes, firm sizes, executive pay, income, and wealth.[16]

[15] Bouchaud [6], Farmer and Geanakoplos [21], Gabaix [24].

[16] Gabaix [23].

Some authors go further stating that those power laws and the resulting exponential scaling are in fact a fundamental characteristic or any structure organized as a network.[17] Moreover, scientists like Geoffrey West and his colleagues from the Santa Fe Institute have endeavored on a research program to investigate the universality of power law distributions and calculate their exponents. Surprisingly, the exponents of similar phenomena are practically the same, hinting at underlying fundamental laws that drive those phenomena.[18] A useful approach to characterizing such phenomena is by taking recourse to the family of power law statistical distributions. The exponential distribution, which is a special type of the Poisson distribution, is an apt choice for such problems.

The distribution is formally defined in Eq. (2.5)[19] where one can clearly observe that its most important parameter is λ—the rate at which a given event occurs. The larger the rate, the steeper the slope of the distribution, and thus—the more unequal the underlying phenomenon.

$$f(x|\lambda) = \lambda e^{-\lambda x} \tag{2.5}$$

This is a particularly approach for modeling events that are not clearly bounded by physical laws which imply clustering around a central tendency but rather allow unbounded growth for some time. Eventually, if the exponent is positive and the growth remains unchecked, the realizations will tend towards an explosive trajectory. This fast growth has often been lauded as something positive and the term "scaling" has entered popular discourse. It is only through scaling that small startups can become the next industry leaders, IT infrastructures can meet ever-expanding demand, and research hubs can provide the needed ever-increasing streams of innovation. In terms of risk, the most interesting region of this distribution is its right tail—there lie the rare events with potential huge impacts.[20] Alternatively, the tail is where the most influential and systemically important organizations and individuals lie. A power law distribution is also very typical of phenomena in the digital economy and is known to characterize popularity and usage patterns in the online world.[21]

The underlying dynamics of a exponentially-distributed variable of interest may be driven by a number of factors. These most likely include one or more of the following[22]:

- **Random growth**—exponential distributions result if there are events driven by random growth such as new realizations of a variable y_{t+1} are dependent on the

[17] West [61].

[18] Ibid.

[19] Due to its form, this equation is defined only for non-negative values of x.

[20] Some readers may be more familiar with the whale curve, depicting the same idea with a concave increasing function.

[21] Johnson et al. [32], Thurner et al. [58].

[22] Gabaix [24], West [61].

present ones y_t with some degree of proportionality encoded in the random variable γ, i.e. $y_{t+1} = \gamma y_t$. Examples of this include population growth, economic growth, or even company growth.

- **Behavioral traits**—peculiarities of human behavior and decision-making can be responsible for such outcomes. Examples of such behaviors may include preferential attachment, least efforts choices, direct and indirect reciprocity and similar ones. Examples include interactions on social media and the popularity of internet celebrities.
- **Superstar dynamics**—there may be market or social structures that disproportionality award outlying performance. This may lead to small differences in objective ability being greatly amplified by the reward structure and thus result in a power law outcome. Examples of this are the paychecks of executive or professional athletes.
- **Resource Optimization**—the objective need to optimize resource allocation in a network or some other well-defined structure, or as the result of some already defined objective function. Both living organisms and cities need to transport energy (and information) to the farthest reaches and the resulting network tends to follow a power law distribution. The scaling laws of metabolic rates, and those of utility networks, both provide examples of this dynamic.

The family of power law distributions is thus able to characterize a very wide spectrum of extremely dissimilar phenomena that are likely driven by a small common set of factors. The emergence of scaling is understood increasingly better and exponential distributions are turning into important tools for risk managers, especially in the case of rare risks, or when modeling the digital realm.

2.3 Simple Risk Metrics

One approach to thinking about risk is by trying to understand how people conceptualize uncertainty and how they react in such situations. This approach gives rise to the idea of individual-level risk perceptions. In finance, the early dominant approach—as exemplified in Modern Portfolio Theory—was radically different. It thought of risk as an underlying fundamental characteristic of any given asset (e.g. stock) or aggregation of assets (e.g. portfolio or market index).[23] Modern Portfolio Theory (MPT) thus shifted the nexus of risk from the individual to the system, thus externalizing those ideas. Their key insight is deceptively simple. If one starts from the Knightian definition of risk, one cannot avoid the conclusion that risk is the deviation from what one expects to happen, and thus the higher the deviation from the expectation, the higher the risk of this asset must be (irrespective of what investors may think or feel). Operationalizing expectations as a variable's central tendency or best forecast,

[23] Elton et al. [15], Fabozzi et al. [18].

then risk is simply its volatility around this point. This can be further operationalized as the variance of the asset.

The rational investor seeks the best tradeoff between expected return (mean) and assumed risk (variance) and this mean–variance optimization is thus the cornerstone of MPT. Individual risk perception and tolerance only as an afterthought when the investor has to select their preferred portfolio out of a pre-approved set of the best ones—the so-called "efficient frontier". This approach is thus highly intriguing as it is based on the idea that risk exists objectively in every asset or transaction, it can be precisely measured based on history alone, and can be managed using mathematical tools towards points of optimality. The externalization or risk, its standardization and its ubiquity at both individual and systemic level put deep marks on risk management thinking for decades to come. Some of those insights are still relevant albeit with significant qualification. This section investigates those ideas in more depth by defining traditional classic risk metrics stemming from finance and looking at their assumptions and implications.

2.3.1 Measures of Variance

When one defines risk as volatility, the most natural way to measure it is through variance—a metric showing how far realization of a random variable (e.g. return of a stock) tend to be from the distribution's peak. Assuming a random distribution as MPT often did, the question collapses to finding two parameters—what the distribution's mean is, and what is its variance or standard deviation. The variance σ^2 captures this volatility and is defined in Eq. (2.8) where x_i is the i-th observation in sample of size n that has a mean of μ.

$$\sigma^2 = \frac{\sum_{i=1}^{n}(x_i - \mu)^2}{n - 1} \tag{2.6}$$

The standard deviation σ is just the square root of the variance. Intuitively, it shows how far is the average observation away from the observed sample mean, μ. Increasing the size of the standard deviation also changes the form of the bell curve as can be seen in Fig. 2.1. This metric is simple and versatile—the standard deviation is easy to calculate, and relatively easy to understand. It preserves the original scale of the variable, e.g. percent, enabling the analyst to sell that a security has a return of 3% with a sigma of 1%. In the case of a well-behaved Normal distribution, statistics gives one a very precise idea of what deviations to expect. If the variable is Gaussian, then 68.1% of all observations fall in the interval $\mu \pm 1\sigma$, 95.4% of all observation are in the interval $\mu \pm 2\sigma$, and the interval $\mu \pm 3\sigma$ contains 99.6% of all observations.

Even if the distribution is not normal, one can use the Chebyshev inequality to get a fairly precise measure of the risk. It essentially states that in any distribution with a mean of μ and a standard deviation of σ, at least $1 - 1/k^2$ observations are within k standard deviations from the mean, or:

$$P(|X - \mu| \leq k\sigma) \geq \frac{1}{k^2} \tag{2.7}$$

The precise conceptualization and, more importantly, measurement of risk held great sway in economics and business ever since the publication of Markowitz's classic article.[24] Some were surprised as it seemed it literally came out of nowhere but its influence on the modern perception of risk cannot be understated.[25] It brought a rigorous understanding of how to manage uncertainty, but the high price economists paid for that were a number of treacherous assumptions. First, MPT largely conceptualized risk as an in-sample property of a group of observations. The risk of the asset is judged by the historical data on it, and the length of the time series is a crucial determinant of the breadth of obtainable knowledge. In reality, some plausible risky events may not have materialized yet but can be plausibly expected (e.g. a nuclear disaster such as the one in the Fukushima Daiichi power plant in 2011). More strikingly, applied analysts often use shorter time series than what is available or focus on more recent events, thus underestimating past but realistic ones (e.g. the Great Depression).

Second, risk is seen as an objective quantity than can be disaggregated from what individuals perceive. It is thus easier to measure and work with from a mathematical standpoint but large parts of the complexity of the phenomenon are lost in this. Individuals may only have preferences on what levels of risk they are willing to undertake but not what dimensions they may be interested in avoiding. While not explicitly said, a standard deviation implies impersonal uni-dimensionality of the concept that leads to a very narrow understanding of risk. Third, using simple MPT-inspired metrics such as the standard deviation precludes the analyst from studying new and emergent phenomena that can only be described with limited or qualitative data. Furthermore, changes in volatility may be only a part of the overall risk landscape and may fail to do full justice to tectonic shifts in the economy. A case in point would be the emergence of Google as a global search engine and data dealer. Looking at volatility tells a poor story of the emergence of the digital economy and the demise of old business models.

Finally, and most importantly, variance metrics fail to incorporate a broader understanding of how an asset is positioned in the overall market and what are the risk drivers that lead to certain volatility. A useful analogy here may a kid looking at a watch—the child does see its hands moving but has little understanding of the mechanics behind that, and also a limited idea of why this is of significance to the adults in the room. Understanding that such a conceptualization of risk is rather wanting, MPT incorporated a number of more sophisticated models that take into account the overall market, structural links between assets and industries, as well as attempts to move to a more forward-looking understanding of risk.

[24] Markowitz [39].

[25] Rubinstein [52].

2.3.2 Market Beta

The question of asset valuation often takes center stage in the discussion of financial risk. Some measure of the reward associated with holding a given asset (e.g. its return) must clearly be associated with the level of risk that this asset holds. This stems from a simple arbitrage argument—rational investors will not hold an asset that entails a higher risk for the same level of return as another asset with lower risk. Savvy market sleuths thus need to be compensated and rewarded for taking up risk. Thus, the dominated asset (the one with higher risk for the same return) should not exist. Conversely, if it did exist, then it must give a higher level of return. This train of thought leads to a simple insight—the price and return must be correlated with the volatility of this asset with respect to the volatility of other assets. In short, the risk premium must be a function of excess risk over the market. The Capital Asset Pricing Model (CAPM) was probably the first model to capture this insight in a rigorous mathematical form, and it left its mark over many subsequent attempts to conceptualize risk in economics and business.[26]

As with many early models, the CAPM is deeply rooted in rational choice theory, based on utility optimization and the resulting competitive market structure. This imposes a number of strict assumptions for its main results to hold. Most notably, the model rues out market frictions (such as delays or transactions costs) and assumes rational utility-maximizing agents that have complete market information. Naturally, this is hardly a realistic description of real-world financial markets but the value and utility of the CAPM stems both from its key insights that are realistic, and also from its approach and thinking about asset risk.

It its simplest form, often operationalized as a regression equation, the CAPM aims to model the excess return (or risk premium) of a given asset. This is defined as the additional expected return ($E[r_i]$) over a riskless rate of return r_f. The risk premium is determined by the volatility of this asset with respect to the excess market return ($E[r_M] - r_f$)—its beta (β_i). Mathematically, this takes the following form (Eq. 2.8):

$$E[r_i] - r_f = \beta_i\big(E[r_M] - r_f\big) \tag{2.8}$$

By rearranging Eq. (2.8) we can also reach an expression that gives an idea of what the asset return should be given three key parameters—the riskless rate, the market return, and the asset beta (Eq. 2.9). The rational investor can then have a feeling of what is an appropriate return for a given level of risk.

$$E[r_i] = r_f + \beta_i\big(E[r_M] - r_f\big) \tag{2.9}$$

One can further operationalize Eq. (2.9) by assuming that the riskless rate at a given point is relatively constant and thus substituting $r_f(1 - \beta_i)$ for β_0 in it and adding an error term u. This is then a simple linear equation that can be estimated by means of Ordinary Least Squares (OLS).

[26] Fama and French [19], Perold [46].

$$r_i = \beta_0 + \beta_i r_M + u \tag{2.10}$$

Using Eq. (2.10), the investor can easily calculate the relevant market beta. If using the linear regression framework, the beta is simply the covariance of the asset's return with respect to the market's divided by the asset's variance (Eq. 2.11).

$$\beta_i = \frac{Cov[r, r_M]}{Var[r_M]} \tag{2.11}$$

The simplicity of these equations is partly responsible for the huge sway they held over finance and business over the last decades. A clearly understood and contextualized metric of risk (the beta) could be easily calculated using relatively simple methods on clearly defined data. This enabled investors to gain a holistic perspective on the individual risks of specific assets as well as to correlate them with overall market conditions. On the one hand, this provided for an expanded and more detailed view of risk. On the other hand, it also boosted investors' confidence in making rigorous financial decisions. More provocatively, it brough large scale data into investment, thus making it one of the first industries to embrace digital technologies and the digital mindset in pursuit of profit.

The CAPM itself is an imperfect approach to viewing total risk and to understanding the functioning of the financial markets, and many criticisms have been levied against it accordingly.[27] Most notably, its assumptions are easy to challenge and there is now wide agreement that they do not accurately reflect neither the market environment, nor human decision-making in it. Its results therefore are to be interpreted with caution as with any model. Strangely enough, the CAPM draws not only the expected level of scientific doubt but also uncharacteristically emotional evaluations, with Pablo Fernandez publishing a paper[28] outright titled "CAPM: An Absurd Model". Strong feelings towards CAPM are in part engendered by how observers connect it to mainstream economics and the idea of efficient markets that helped promote controversial deregulation.[29] Indeed, it is a pinnacle of rational choice theory and carries many of its problems but is nevertheless an important stepping stone towards better conceptualizing and understanding financial risk.

2.3.3 Risk and Return Tradeoffs

Mean–variance optimization in MPT is all about finding an optimum tradeoff between reward (expected return) and risk (asset volatility). This comes to the heart of taking action under uncertainty. Since it is impossible to gain any meaningful return in the absence of risk, and risk is impossible to fully diversify or eliminate,

---

[27] Abbas et al. [1].

[28] Fernandez [22].

[29] O'Sullivan [42].

then the only reasonable thing for a rational individual to do is to make a risk-adjusted decision and then track risk-adjusted performance. Trivial as this may seem, it may often be the case that a skewed incentive structure can lead to excessive risk taking with short-term benefits but catastrophic long-term results. Researchers[30] have noted that during the run-up to the global financial meltdown of 2007–2008 executives at key investment banks were incentivized and rewarded for taking needlessly large risks that bring profit in the short-run but disaster over the medium one. While this action may be individually rational (executives did retain their earnings despite the calamitous outcome for their organizations), this is hardly desirable at the systemic level.

This brings an idea of how to measure rigorously the risk-return tradeoff in order to support rational decisions. A first pass to such a metric is the Sharpe ratio, S. It aims to show what is the reward achieved per unit of risk undertaken. The Sharpe ratio is usually defined as a fraction where the numerator is the expected risk premium over some riskless rate ($E[r - r_f]$) and the denominator is some measure of risk such as the standard deviation or a value at risk metric (here denoted as σ).

$$S = \frac{E\left[r - r_f\right]}{\sigma} \tag{2.12}$$

While the simple Sharpe ratio may have some problems when used as a single metric in a complex risky environment,[31] it can still serve as an important focal point for personal and organizational decisions in pursuit of risk-adjusted benefits. Once the idea of formally optimizing the tradeoff between reward and volatility take hold, efforts at minimizing overall exposure naturally become imperative. MPT postulates that risks can either be general and non-specific to the asset (so-called market risk) or particular to a given asset (specific risk). The former may include events such as an economic downturn or a pandemic that exert a negative influence on practically all assets.

The latter reflect specifics such as the management behind a stock's company, the competition it faces, or its mastery of technology and innovation. Those are unique to an asset and can thus be diversified away by creating a portfolio of dissimilar assets. Diversification of risk proved to be a powerful and resilient idea and has found its way not only in financial and business risk management but also in diverse decisions ranging from the personal (e.g. the acquisition of transferable skills) to the global (e.g. the development of multiple vaccines during the coronavirus pandemic in 2020–2021). One may say that this concept is close to folk wisdom advising never to have all eggs in one basket but risk management does provide a more sophisticated and markedly quantitative treatment to this idea.

Diversification can be aptly illustrated with a simple example of a portfolio containing two assets. Their returns are denoted as r_1 and r_2, and their weights in the portfolio—w_1 and w_2, respectively. The total portfolio return, r_P, is given by

---

30 Bebchuk et al. [4].

31 Farinelli et al. [20].

the following equation:

$$r_P = w_1 r_1 + w_2 r_2 \tag{2.13}$$

The total portfolio risk (standard deviation, σ_P) is defined by the individual risks (σ_1 and σ_2), weighted by the fraction that this asset takes in the portfolio. However, the portfolio risk also depends on whether these assets tend to move together (i.e. they both rise or fall in value simultaneously) or in opposite directions (when one gains value, the other one loses) and the strength of this association as measured by the correlation coefficient ρ. Thus, total portfolio risk is given by the following equation:

$$\sigma_P = \sqrt{(w_1\sigma_1)^2 + (w_2\sigma_2)^2 + 2\rho(w_1\sigma_1)(w_2\sigma_2)} \tag{2.14}$$

It is the negative asset correlations that make diversification possible. In the very extreme case when the asset returns are perfect opposites of one another, then $\rho = -1$, and thus Eq. (2.14) will reduce to the following form:

$$\begin{aligned}\sigma_P &= \sqrt{(w_1\sigma_1)^2 + (w_2\sigma_2)^2 - 2(w_1\sigma_1)(w_2\sigma_2)} \\ &= \sqrt{(w_1\sigma_1 - w_2\sigma_2)^2} = w_1\sigma_1 - w_2\sigma_2 \\ &= w_1\sigma_1 - (1 - w_1)\sigma_2 \end{aligned} \tag{2.15}$$

In this case, one can completely eliminate portfolio risk by setting the assets weights to be exactly proportional to their individual risks (Eq. 2.16). This shrewd selection of weights ensures that any loss in one asset will be exactly and perfectly offset by a gain in the other. On the darker side, profit is impossible as any gains are offset by equal in size losses.

$$\sigma_P = 0 \Rightarrow \frac{w_1}{1 - w_1} = \frac{\sigma_2}{\sigma_1} \tag{2.16}$$

In the real world, one almost never finds perfect correlations which puts practical obstacles to perfect diversification. However, sizable negative correlations do exist and enable researchers and practitioners to design balanced and diversified portfolios where risk is minimized by effective bundling of different assets. The estimation of both risk metrics and the correlation patterns depends on past data and is not guaranteed to be time-invariant or to hold in the future. Particularly in the case when a rare high-impact event occurs, possibly causing a regime switch, the correlational structure of assets can change so drastically that it eliminates some (or all) of the benefits of diversification. Still, the realization of risk-return tradeoffs together with diversification created a more nuanced understanding of risk management and paved the way towards both using more advanced methods as well as leveraging a hedging strategy for risks that matter.

2.4 Advanced Risk Metrics

Simpler variance metrics for risk tend to be parameters whose point parameters are estimated over a certain sample. In this sense, indicators like the standard deviation are merely a snapshot of the analyst's best guess of risk given available data. On the other hand, a risky variable can be much better characterized not by such a point estimate but rather by taking recourse to its complete statistical distribution. Having this under the hood, a risk manager can precisely state what are the probabilities of incurring a negative event with a given magnitude (e.g. there is a 90% chance that losses will not exceed 5%). So the natural evolution of risk metrics proceeded to look into overall probability distribution of the variables of interest and draw inferences from those. From this, two key metrics were born—the Value at Risk (VaR), and the Expected Tail Loss (ETL).[32]

2.4.1 *Value at Risk*

The Value at Risk metric aims to answer a straightforward question: given a random variable of interest (e.g. asset returns), what is the maximum expected loss that can be incurred over some time interval with a certain level of probability.[33] A VaR metric can thus state that over the next week an organization expects a maximum loss of less than 4% on its portfolio of assets in 95% of the cases. This confidence interval is usually denoted as alpha (α), and the whole metric—as VaR_α. Further denoting the expected loss as L, and the loss actually realized as l, then VaR_α is defined as follows:

$$VaR_\alpha(L) = \inf\{l \in \mathbb{R} : P(L > l) \leq 1 - \alpha\} \tag{2.17}$$

If one assumes that the L follows a well-defined distribution function F_L, then the definition in Eq. (2.17) can be expressed as follows:

$$VaR_\alpha(L) = \inf\{l \in \mathbb{R} : F_L(l) > \alpha\} \tag{2.18}$$

The logic behind the Value at Risk at a certain level (e.g. 95%) is schematically presented in Fig. 2.6. Initially the analyst constructs the variable's distribution and then estimates what is the number that is smaller than 95% of all realizations. This is the $VaR_{0.95}$ and its shows what is the largest expected loss in 95% of the cases (the shaded region in Fig. 2.6). Thus, the daily VaR at 95% gives an idea of what is biggest possible loss in 19 out of 20 days. The risk manager may wish to impose a more strenuous level for the daily metric, e.g. 99%, thus measuring the maximum loss that will be incurred in 99 out of 100 days. Typically, this is larger than the $VaR_{0.95}$ as it captures a grater part of the distribution's tail.

---

[32] Chen [9].

[33] Esterhuysen et al. [17], Jorion [33].

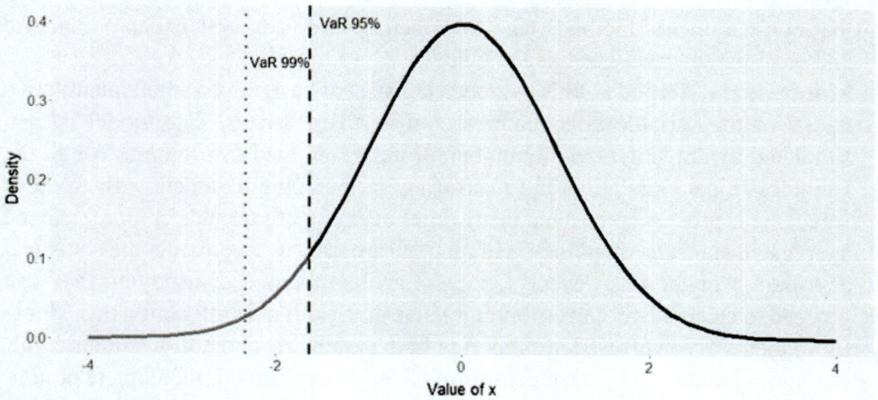

Fig. 2.6 Value at Risk at 95% and 99% level

The key problem in estimating the VaR metric is finding the actual underlying distribution of the variable under study. This can be done by relying on empirical data, inspecting its characteristics and then estimating the parameters of the most likely distribution. The standard default is to leverage a Gaussian distribution and numerous statistical tests can be performed to investigate whether it is a good fit. Risk managers, however, must beware as sometimes those tests lack sufficient power to reject the null hypothesis of normality and so erroneously accept it.

In the final analysis, there are three major approaches to calculating VaR metrics[34]:

- **Parametric Estimation**—this approach leverages historical data to estimate both the type and the parameters of the statistical distribution. Crucially, it makes assumptions on the distribution type (e.g. a Normal one) and proceeds to estimating its parameters through some sort of optimization algorithm. The parametric estimation needs a relatively long time series to ensure stable and unbiased estimates and is rather sensitive to changing the type of underlying distribution assumed. Some modifications may be used to correct for possible non-normality or skewness of data such as the Cornish-Fisher Method (CFM).[35]
- **Non-parametric Estimation**—this approach uses historical data as well but, crucially, does not make assumptions on the type of underlying distribution. Sometimes those approaches may use a simulation drawing upon existing data realizations and resampling them, thus forming new scenarios. Overall, the statistical properties and the values of these scenarios are used to estimate the Value at Risk and possibly other relevant metrics. The key benefit of non-parametric approaches is that it practically avoid the issue of model misspecification and the questionable parameter estimates stemming from. There is mounting research

[34] Guharay et al. [27], Koike and Hofert [34].

[35] Kokoris et al. [35].

that non-parametric methods tend to sometimes perform better than parametric and semi-parametric ones like the CFM.[36]

- **Monte Carlo Methods**—this approach consists of designing a small quantitative model of the variable under interest and its main drivers. The model is then simulated by drawing random numbers from pre-defined distributions for all the variables in the simulation and measuring the outcome parameters given those values. This process is repeated many times over (e.g., thousands or more) and the aggregated statistics of all those simulations are used to reconstruct the statistical distribution of the risky event. Once the analyst has the distribution, they can proceed to calculate its corresponding Value at Risk. It is worth noting that Monte Carlo methods can be used in support of both parametric and non-parametric VaR estimates, but they are extremely useful when data is limited, incomplete or non-existent. Thus, they are a versatile tool to aid both calculation and imagination when dealing with risk.

As rich and extensive as it is, the VaR metric has not only been the focus of a large body of research but has also seen extensive practical applications ranging from estimating risk in financial markets, through medical and engineering risk, all the way into different operational risks. This wide acceptance and adoption hinges on its many strengths.[37] Notable among them are the following:

- Provides a common, consistent and integrated measure of risk—VaR is well-understood and can be applied across a diverse set of businesses, activities, processes and assets. The potential loss of every single one of those is expressed in meaningful business terms that can aid organizational decision-making. It also provides for an integrated overview of risk that can be extended from the very detailed level (e.g. VaR of a given asset over the next week) to a more topline overview (total operational losses for the next year). An additional benefit here is that if assumptions and methodology remain invariant, the VaR estimates will be deterministic.
- Provides an aggregated measure of risk and risk-adjusted performance—the VaR metric provides not only an integrated view of risk, but more importantly—a measure of risk-adjusted performance (e.g. what return was achieved given the risk undertaken). This can guide management decision-making and also be used to map and improve the incentive structure across divisions, processes and individual agents. Using Value at Risk metrics also enables organizations to assess and compare different portfolios and actions and thus optimize not only their tactical approach but also their strategic goals, ensuring that optimal risk-return tradeoffs are achieved across the full spectrum of activity.
- Easy to communicate and comprehend—the Value at Risk metric has the additional advantage that it is intuitive to explain and understand, thus greatly aiding communication. Despite the fact that some regulators are moving away from the VaR metric, it has long been required as part of compliance and is thus familiar and

[36] Huang et al. [30].

[37] Crouhy et al. [12].

accepted. Additionally VaR compresses overall risk (of organization, unit, portfolio, position, asset or activity) into a single number thus stripping the complexity of risk exposure. This may be useful for practical decision-making across groups of stakeholders with diverse knowledge, expectations, and risk awareness.

Despite its many benefits, VaR has often been the subject of sustained and harsh criticism. Its key drawbacks are that the large number of assumptions required, the diverse estimation methodologies and interpretation issues make it an imperfect measure of risk.[38] First, parametric and semi-parametric VaR calculations take recourse to the assumption that the data under investigation follows a well-defined (often Normal) distribution which may not be met in practice. The metric further assumes that market conditions under which data was collected will persevere, and that the fundamental risk of assets or activities remain time-invariant. Should these hold, the VaR calculation produces meaningful estimates for the future; should these fail—results are unsavory. Second, a large focus of VaR metrics falls on quantifiable and measured risks for which sufficient time series data are available. In practice, many qualitative or rare risks may dominate the risk exposure at given times (e.g. change in government or global pandemic). Those are precisely Taleb's black swans that VaR is rather powerless against.

Third, using VaR metrics may give a false sense of precision and confidence. The VaR methodology produces precise quantitative estimates that may give false comfort to decision-makers that they are dealing with precise estimates of risk instead of the rough guidelines they actually have. Such false confidence may result in actions that worsen instead of improve the risk profile and exposure of individuals and organizations. Fourth, VaR metrics gives a rather imprecise estimate of expected maximum loss as in practice the realization may well exceed it. This true by the very definition of the metric—as 95% VaR will give the maximum loss in 95% of the cases. Conversely, in 5% of the cases the loss can be much larger—and it is precisely those catastrophic events that truly shape the risk exposure.

Taking all those weaknesses into account, Wallstreet pundits half-jokingly argue that VaR is just like an airbag that functions every single time except when you need it. This sentiment captures the notion that VaR works best in stable and predictable environments where no large shifts (regime changes) occur. In a realistic and complex world such a metric suitable for stable environments can fail spectacularly in case rare occurrences happen, while at the same time giving decision-makers and stakeholder a false sense of understanding and control. Precisely in this vein, some researchers have argued that risk metrics such as the Value at Risk have contributed to the global financial meltdown of 2007–2008.[39] Those weaknesses have also spurred regulators to move away from Value at Risk and propose a new integrated risk adjusted metric for financial institutions instead.[40]

[38] Chen [9], Yamai and Yoshiba [62].

[39] Sollis [56].

[40] Wang and Zitikis [60].

2.4.2 Expected Shortfall

The VaR metric is a risk measure for ordinary days. It purports to calculate expected loss in a large percent of cases (e.g. $\alpha = 95\%$ or even $\alpha = 99\%$) and thus help guide decision-making. However, most individual and organizations' risk exposure are not dominated by well-understood and managed risks that happen all of the time—rather, they are crucially shaped by large and unexpected occurrences with potentially catastrophic impact. In short, it is what Nassim Taleb called "black swans" are what tends to matter the most. The question of their magnitude is thus equally, if not more important, than that of the ordinary risks.

Measuring Value at Risk provides a rough guideline to what is expected most of the time in α cases (i.e. the worst possible loss in 95% of the cases). An alternative metric—the Expected Shortfall (ES) gives indication of what the expected loss will be in the other $1 - \alpha$ cases. ES thus aims to provide insight of the catastrophic rare losses that an organization may expect and since it investigates the extremes (or tails) of statistical distribution, it is also known under the name of Expected Tail Loss (ETL). The Expected Shortfall at level α is denoted as ES_α, and thus the tail loss at $\alpha = 5\%$ is written out as $ES_{0.05}$. Thus the combination of ES_α and $VAR_{1\text{-}\alpha}$ allows the decision-maker to gain a more fuller understanding of the total risk exposure across the whole distribution. ES_α is defined as follows (with $\gamma = 1 - \alpha$):

$$ES_\alpha = \frac{1}{\alpha} \int_0^\alpha VaR_\gamma(X) d\gamma \tag{2.19}$$

Equation (2.19) clearly underlines the fact that the ES metric is typically larger in magnitude than VaR, and this holds particularly true if the data follows a fat-tailed distribution. In terms of calculation, the Expected Shortfall can be calculated using the same methods and approaches as the Value at Risk—the parametric and semi-parametric ones, the non-parametric ones, and the Monte Carlo simulations (or a combination of those). One should note that since the focus of ES is the tail loss, and since typically observations from the tail of the statistical distribution are a small fraction of samples under study, the calculation of the ES_α generally requires larger samples than that of $VaR_{1\text{-}\alpha}$.[41]

The basic idea behind the Expected Shortfall metric is presented graphically in Fig. 2.7. Once the analyst has constructed the probability distribution of the variable under interest (e.g. asset returns), one can estimate the precise number above which the realizations of the random variable will fall 99% of the time. All the values of x that are below this number form the region which the Expected Shortfall at 1% seeks to model (shaded area in Fig. 2.7). Calculating the mathematical expectation of those numbers produces the $ES_{0.01}$—the expected tail loss in 1% of the cases. Decision-makers may instead opt for a less aggressive estimate of the heavy loss

[41] Yamai and Yoshiba [62].

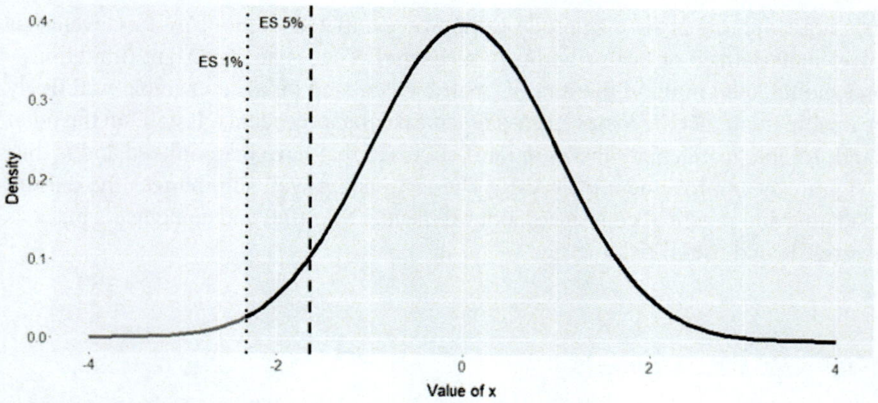

Fig. 2.7 Expected shortfall at 1% and 5% level

and calculate the $ES_{0.05}$, which will typically represent smaller expected loss that the $ES_{0.01}$.

The Expected Shortfall is thus a more conservative metric and very suitable for situations with large amount of uncertainty. In addition to this, the Expected Shortfall, unlike VaR is a coherent risk measure, leading to the same consistent results when applied to continuous distributions and can also be estimated effectively in cases in which the usual VaR estimators fail.[42] It is hardly a surprise that both the academic discourse and regulatory practice are shifting away from using VaR and showing a clear preference for ES instead.[43] While much improved, the Expected Shortfall metric may also be thought (with admittedly some simplification) as the mirror image of Value at Risk, covering all that VaR has left behind. This enables ES to sport the strengths of VaR but also to be vulnerable to its weaknesses.

In terms of strengths, both metrics are relatively easy to understand and communicate while providing a single integrated and standardized measure of risk across organizations. They are both recognized by experts and regulators and enable analysts to obtain a comprehensive view of the organization's full risk exposure, to compare units, processes, and activities, and optimize across them. However, just like VaR, Expected Shortfall has some weaknesses. First, in its parametric and semi-parametric estimations, it remains highly sensitive to the assumptions for the underlying distribution. Changing this assumption may lead to large quantitative differences that skew decision-making. Second, ES may breed as much confidence as VaR, if not more. This false sense of safety must be understood and proactively fought.

Third, ES also focuses on risks that can be quantitatively measured and skirts the issue of more intangible risks and uncertainties which resist quantification. This is particularly onerous for the expected tail loss, as it is the rare and unexpected events that may dominate tail risk. Essentially the analyst calculates a metric for

[42] Acerb and Tasche [2].

[43] Chang et al. [8].

the catastrophic, but its magnitude cannot be estimated as the truly rare event has never materialized or been rigorously measured. That is an important limitation—one should keep in mind that the ES metric calculated on a data sample will likely be unable to predict the losses in case of catastrophic precedents. It will, on the other hand, be able to calculate losses in unlikely event that have materialized and is thus a significant improvement over VaR. The last insight well summarizes the current academic and practical consensus about Expected Shortfall and cements its place as a versatile and useful risk metric.

2.4.3 *Risk as Anomaly*

Both the Value at Risk and the Expected Shortfall metrics focus heavily on the underlying statistical distribution of the variable under interest (e.g. returns). This enables them to provide rigorous and consistent answers to all probabilistic questions of either losses (downside risks) or gains (upside risks), thus fully characterizing the organizational risk exposure given the data sample used. This versatility comes at the high price of having to assume or construct the probability distribution—an exercise that can be messy, rough, and ridden with untenable assumptions. Moreover, in some applications this level of precision is hardly needed as a binary metric (risky or not) may suffice for rational decision-makers to take action.

This train of thought leads to the possibility of conceptualizing risk in a different manner. Defining risk as the deviation from a given expectation, it is clear that the greater this deviation is, the greater the risk. Consequently, highly deviant and anomalous observations within a given sample are very likely realizations of risky events. This simple observation allows one to rely on a purely data-driven definition of risk—risks are sufficiently anomalous observations in a dataset. Those anomalies are relatively rare and thus when plotted on the statistical distribution of the data they fall exactly in its tails. Thus, defining risk as an anomaly, the analyst looks at the same underlying phenomenon as with VaR and ES (tail observations) but from a markedly different lens. On the one hand, this lens is somewhat less precise but on the other it frees the decision-maker from the burden of making quite a few unrealistic assumptions.

This intuition is presented graphically in Fig. 2.8, displaying a random sample, drawn from a Normal distribution with mean 0 and standard deviation of 1, *N(0, 1)*. The overwhelming majority of observations are common and legitimate—they cluster in the middle of both distributions and so are expected with a low level of risk. Graphically, they form a relatively large and homogenous cluster. Those observations at the tails of the distribution are the anomalous, and thus potentially risky, ones. They are visually separated from the rest and hint at potential edge cases in real life that can have crucial implications for the overall risk exposure. In this view, risk management is all about identifying and addressing those anomalies. Following this approach, research has focused on detecting the risks of fraud, IT risks, modeling

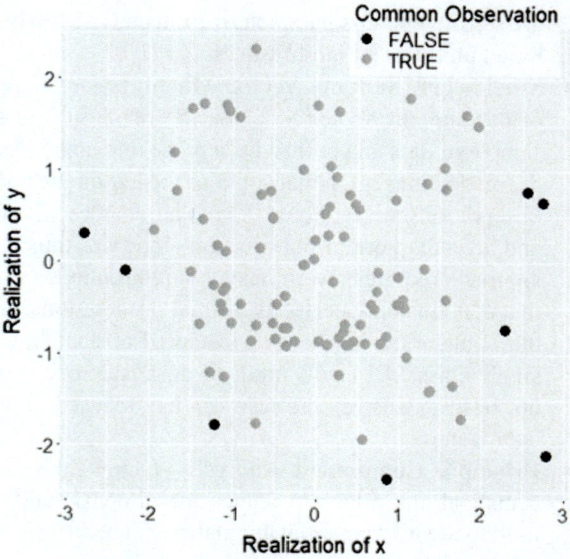

Fig. 2.8 Common and anomalous observations in a random sample as per DBSCAN algorithm

online behavior, identifying medical risks, as well as many applications in the Internet of Things (IoT) domain.[44]

The key question for understanding risk from this perspective is all about being able to identify the anomalies using a set of consistent criteria, methodologies, and algorithms. These may vary from the very simple to the rather sophisticated, and some of the most widely used ones follow:

- **Clustering methods**—this type of methods assumes that anomalies can be distinguished by dividing the sample into clusters and investigating their structure. Normal and legitimate observations are supposed to form large and relatively homogenous clusters, while risky observations form small and distant clusters or do not belong to any cluster. The key idea here is how far is a given observation from the standard one. The analyst can estimate the distance (using a distance metric) or see if the observation is within a well-population neighborhood or not (density-based clustering). The simplest clustering algorithm is k-Means but more sophisticated ones such as the DBSCAN[45] or the Local Outlier Factor (LOF)[46] have gained traction and prominence. Density-based algorithms such as the latter two are both powerful and versatile but suffer from a significant drawback—the user must input parameters than characterize the size and composition of the neighborhood that is then used to evaluate anomalies. Both DBSCAN and LOF

44 Chandola et al. [7], Habeeb et al. [28], Omar et al. [43], Phua et al. [47], Qiu et al. [48], Rousseeuw and Hubert [50].

45 Schubert et al. [54].

46 Gogoi et al. [25], Mishra and Chawla [41].

are sensitive to this parameterization and are likely to yield very different results based on different initial values.

- **Single-class Support Vector Machines**—this approach builds upon the logic of the Support Vector Machine (SVM) family of algorithms. It takes an initial sample of data and constructs a plane that contains all initial data points which is deemed as the region that contains the legitimate common observations.[47] As new observations are fed to it, the model decides whether they belong to this region and are thus normal ones, or not—and are thus labeled as anomalies. Such an approach requires a very risky assumption indeed—that the analyst has a sample that contains only perfectly normal observations, and that they completely cover the boundaries of what is considered normal. If either of those is not met, the single-class SVM will produce unsatisfactory results. This is why its use has not been extensive, and research has focused on competing anomaly detection approaches.
- **Principal Component Analysis**—factor analysis is among the most popular statistical algorithms to reduce the dimensionality of a dataset but can also be used to identify potential anomalies.[48] The Principal Component Analysis (PCA) is one of the most common methods in this family. The idea behind it is rather simple. Mapping to a lower dimensional space that the original one, data is divided into different components that aim to capture most variance. The first principal component constructed hold the largest amount of variance, followed by the second, and so on. Thus, the first few principal components describe most of the general and ordinary characteristics of the dataset, while the last ones tend to capture its peculiarities. Every individual observation can then be mapped onto those components and see if it is highly correlated with it (i.e. has a large factor loading) or not. Common observations tend to have high factor loadings on the first principal components, while anomalies tend to have high loadings on the last components representing peculiarities in data. Thus component loadings can serve as a effective measure of how anomalous, or risky, an observation is. Notably, PCA performance is highly dependent on the underlying data structure.
- **Statistical Methods**—a large number of familiar statistical methods can be leveraged to detect extreme observations (anomalies).[49] Initially designed to identify and help address outliers in data, today these methods are used for risk management, as well. The simplest ones include using visualizations and simple metrics such as distance from a given interquartile range or the mean. More sophisticated methods include calculating distance measures from a reference point such as the Mahalanobis distance and formally testing whether some group of anomalies are statistically significantly different from a set of normal observations, using a t-test, a chi-squared test, or a similar statistic. Finally, a number of regression approaches may be used to identify anomalies. When estimating a regression model, one can calculate a measure of the model errors (difference between actual and fitted value)

---

[47] Erfani et al. [14].

[48] Child [11].

[49] Chandola et al. [7], Rousseeuw and Hubert [50].

and use each observation's error as its risk score. A particularly popular task is to define anomalies in a given time series by decomposing the series and identifying values that are notably different. This can then be applied to practical problems such as fraud detection, identifying change of market conditions or dealing with unexpected spikes in demand. Statistical approaches have the benefit that they are well-understood and based on a rigorous foundation that enables one to pinpoint the uncertainty of the estimates through confidence intervals. On the other hand their assumptions may not be met in practice, thus skewing results.

In the final analysis, treating risk as a set of anomalous observations rather than as a random variable with well-defined underlying statistical distribution is merely another way of conceptualizing uncertainty. Either way, the analyst focuses on the tails, where the unexpected lurks, and is able to identify risks that lend themselves to management. In research and practice, the former view of risk as a random variable tends to be more associated with financial risks, critical infrastructure risk, individual choice, and market conditions uncertainty. On the other hand, focusing on anomalies is the typical way to go forward in fraud detection, credit risk, and has application in operational risk as well.

2.5 Empirical Stability of Metrics

Risk is a basic feature of individual activities, processes, agents, or assets. Each of those has an objective level of risk, driven mostly by its fundamentals and context. This fundamental risk is hidden from even the shrewdest of analysts and the best one can hope for is to devise a metric or proxy that tries to capture it. Measurement of the Gross Domestic Product (GDP) may be an apt analogy here. While we may want to capture the total level of economic activity—a fundamental objectively existing characteristic of each country—we cannot do so, as it is very difficult to observe and quantity each transaction. Instead, economists use the sum of value added in non-household non-criminal production of goods and services (or the GDP) as an approximation of all activity. While much richness is certainly missed, this proxy may be sufficient for practical purposes. In a similar vein, the risk analyst calculates quantitative metrics that merely approximate the levels of objective risk but may still be useful for making decisions and managing uncertainty in a complex world. Naturally, this data is passed through the lens of individual judgment, thus reaching a final subjective evaluation of risk.

2.5.1 Simple Risk Metrics

Simple risk metrics have the benefit of being easy to understand and communicate which makes them a useful first pass at understanding risk. Most importantly they are also a deterministic calculation of asset, agent or activity features that yields consistent numbers irrespective of assumptions and calculation approaches. The three most common basic metrics reflect the following:

- **Volatility**—how far away from a given well-defined expectation (e.g. the mathematical expectation) may the realizations of the random process fall. This is usually measured through the variance or the standard deviation (see Eq. 2.6).
- **Reaction to market**—showing how a specific asset reacts to market dynamics, pointing at its diversification potential. The asset beta in Eq. (2.11) is the most common metric for that.
- **Rist-return tradeoff**—this shows the amount of return that is given to the investor for assuming a certain amount of risk. Different versions of the Sharpe ratio (see Eq. 2.12) are used to this end. The denominator—i.e. the risk measure—is most often the standard deviation, but other metrics such as the VaR or the ETL can be used.

Acknowledging that irrespective of the metric used, a higher fundamental level of risk should bring a higher expected return, one can further analyze the risk-return tradeoff. Using daily returns data on 25 European stock markets over the period 2006–2020 (see Table 2.1 for a list) it is straightforward to calculate their expected returns (long-run mean) and to proxy their risk level through registered volatility (i.e. the standard deviation). Results are shown in Fig. 2.9, which displays the clear positive correlation between the expected returns and the observed level of risk. It is clear that the arbitrage condition is largely responsible for this. It broadly states that an asset with a certain risk level will only be held by investors if it offers a better return than a comparable asset with a lower risk. Should that not be the case, investor will find it irrational to buy into the dominated asset, thus driving it out of the market. This condition is better interpreted as the long-run direction of the market rather than an equilibrium condition that holds in every single moment.

While it is perfectly true that rational investors make this optimization decision, driving down the demand of dominated assets it may still be possible to find assets or even groups of assets (as proxied by the market index) that are not optimal. The trend line in Fig. 2.9 gives the average tradeoff between the two—with a $\beta = 0.36$ (and a $p = 0.00012$), it means that investors generally require 0.36% higher returns for taking up a 1% increase in standard deviation (volatility). However, indices below the line promise a worse tradeoff, while those above it—a better one. Under the arbitrage argument, underperforming assets in the suboptimal indices should be driven out of the market until the indices are comparable across Europe and one starts to observe similar risk-return tradeoffs. This is not the case for a variety of historical, institutional, regulatory, social, and psychological reasons.

The insight is that while risk has important economic dimensions that are captured by its corresponding level of return or reward, it is also true that those tradeoffs are indicative and are modified by both the decision architecture and the participating agents with their many peculiarities. Thus, it is perfectly possible to find dominated or overvalued assets that are continuously traded or held by investors. On the one hand, this opens the possibility of generating economic profit by betting on fundamentals.

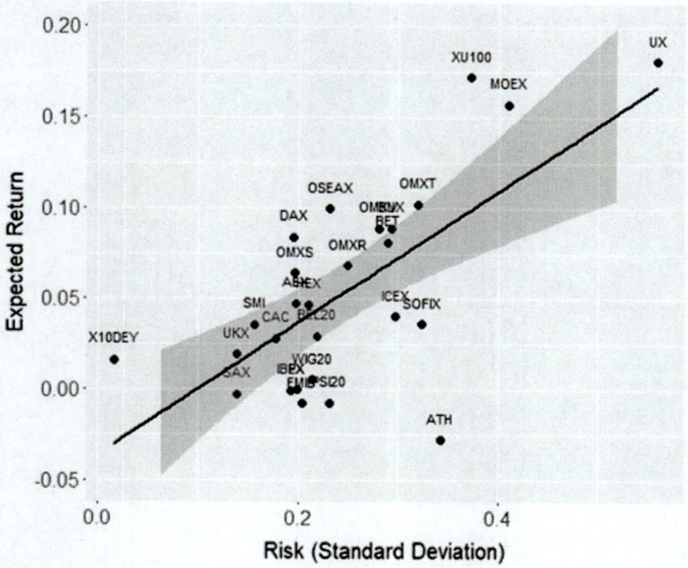

Fig. 2.9 Risk-return tradeoff in selected European stock markets

On the other hand, this is the stuff economic bubbles and their ensuing crises are made of. As Keynes himself aptly observed in this respect: "the stock market can remain irrational longer than you can remain solvent".

Once sufficient data on asset performance is collected, the analyst may want to model in more depth their statistical properties. Those are best summarized in the statistical distribution of the variable under interest. Figure 2.10 presents the daily returns distribution of four European markets over the period 2006–2020. While they all resemble the bell-shaped Normal distribution, the analyst should note that they are not perfectly Gaussian. Thus, relying on the assumption of normality will give only an approximation of the actual underlying distribution. In some instances, this may be enough while in others—woefully inadequate. There are generally two routes when using historical data for risk calculation—assuming that it follows a certain distribution and doing a parametric estimate based on it, or shunning away this assumption and relying on a non-parametric one, instead. Both have their strengths and weaknesses that can be illustrated by calculating two common risk metrics—the Value at Risk (VaR) and the Expected Shortfall (ES).

2.5.2 *Value at Risk*

Using the 15-year time series of European stock market performance, we calculate the daily 95% Value at Risk metric (see Eq. 2.18), using three main approaches:

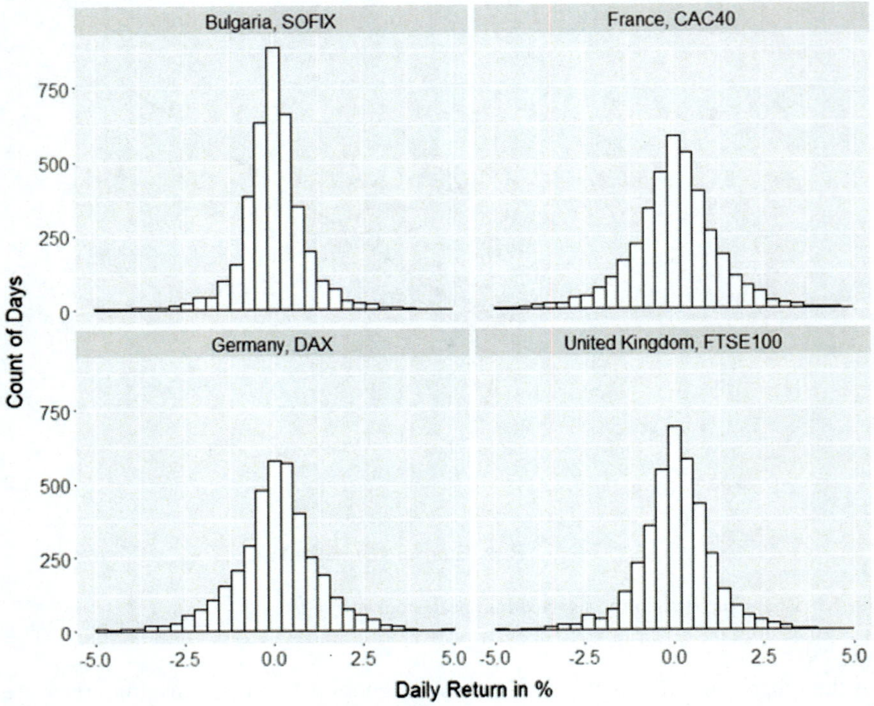

Fig. 2.10 Histogram of daily returns in selected European stock markets

non-parametric historical estimate, parametric estimate assuming a Normal distribution, and a modified VaR that leverage the Cornish-Fisher correction for fat-tailed distributions.[50] Results are presented in Table 2.1. The first thing to observe is that the expected losses (hence risk) in European stocks vary widely across European markets. They range from a low of around 1.5% for markets such as the Baltic stock exchanges and Bulgaria to a high of 2.5% for Italy, Ukraine and Russia. Developed and deep markets with strong fundamentals like the German, Belgian, French and UK stock exchanges cluster around 2%. Thus, in those markets, one should not expect daily losses to exceed 2% in 19 out of 20 days. While the different approaches to VaR calculation seem to point to similar levels of risk, their quantitative estimates may differ—sometimes significantly. In most markets the range of difference between the highest and the lowest estimate is around 10% or less but there are some notable outliers which feature extreme ranges such as the Russian (79%), Latvian (43%) and Lithuanian (36%) indices. It seems that for relatively stable markets with a long time series of data the analyst may choose whichever estimation approaches with little quantitative impact. For markets with a more pronounced dynamics, the choice of method is correspondingly more important.

---

[50] For more details on calculation implementation and specifics, refer to Peterson et al. [45].

The estimation of the Value at Risk metric is particularly problematic but also consequential whenever the asset or activity under study undergoes significant increase in the level of risk and thus volatility. It is precisely here that estimates are less stable, but their accuracy is of greater importance. To investigate further which approach is suitable for stormy times one can compare the accuracy of forecasted to realized VaR for a period of turbulence. In order to do this, 95% daily VaRs are calculated for all markets under study using data from 2006 to 2019 and those numbers are compared to the realized baseline VaRs for 2020—a year of no small volatility. Results of the error rates are shown in Table 2.2.

The absolute error—difference between forecast and realization—is somewhat smaller in the parametric estimate that assumes a normal distribution, but the difference is to a non-parametric one is relatively small. The Gaussian parametric estimate has a significantly lower mean absolute percentage error (MAPE) than both the non-parametric and the modified fat-tailed estimate. There is also little difference in the root mean squared error (RMSE) between the historical and the Gaussian approach with the modified one have a notably higher RMSE. The performance of the modified VaR is somewhat surprising. The Cornish-Fisher correction is applied to take account of possible fat tails of the distribution whereby extreme losses are more likely.

The year 2020 is an apt case in point—it saw the onset of twin economic and public health crisis that deeply affected European markets. It is precisely such unexpected events that push returns into the tails of the distribution. However, the modified VaR numbers were never higher than the ones generated by the parametric Gaussian approach, and thus failed to adequately forecast risk. This shows that in the presence of long time series that encompass at least one full economic cycle, as is the case with the European data under study, the analyst may safely choose to rely on the assumption of normality and do a parametric VaR estimate.

2.5.3 *Expected Shortfall*

The 95% Value at Risk metric answers a simple question: what is the worst that can happen in 95% of the cases. This clearly begs the mirror question—what happens in the rest 5% not covered by the VaR. Pundits sometimes summarize it: when things get bad, how bad do they get? The Expected Shortfall, ES (or Expected Tail Loss, ETL) answers this question. Focusing on extreme cases that usually comprise the most impactful risks is of obvious interest, and the ES metric is slowly taking over as the preferred number for practitioners, regulators, and academic alike. Using the same data on 25 European stock markets, we calculate their respective ETLs (as per Eq. 2.19). Again, the three main approaches used are the historical non-parametric one, the parametric Gaussian one, and the Cornish-Fisher modification.[51] Results are presented in Table 2.3.

[51] Peterson et al. [45].

Table 2.1 Daily 95% Value at Risk (VaR) metrics for selected European stock markets

Country	Exchange	Index	Historical VaR (%)	Gaussian VaR (%)	Modified VaR (%)	Range (%)
Belgium	Euronext Brussels	BEL20	2.03	2.12	2.02	4.76
Bulgaria	Bulgarian Stock Exchange	SOFIX	1.50	1.84	1.82	22.04
Czechia	Prague Stock Exchange	PX	1.97	2.23	1.89	17.94
Estonia	Estonian Stock Exchange	OMXT	1.45	1.68	1.39	20.93
Finland	Helsinki Stock Exchange	HEX	2.15	2.21	2.11	4.46
France	Euronext Paris	CAC40	2.17	2.32	2.11	9.82
Germany	Deustche Boerse	DAX	2.15	2.25	2.03	11.03
Greece	Athens Exchange	ATH	3.20	3.32	3.18	4.41
Hungary	Budapest Stock Exchange	BUX	2.30	2.48	2.26	9.94
Italy	Borsa Italiano	FMIB	2.54	2.65	2.56	4.14
Latvia	Latvian Stock Exchange	OMXR	1.78	2.04	1.43	42.78
Lithuania	Vilnius Stock Exchange	OMXV	1.32	1.64	1.21	35.76
Netherlands	Amsterdam Stock Exchange	AEX	2.02	2.14	1.96	9.51
Norway	Oslo Stock Exchange	OSEAX	2.22	2.33	2.35	5.91
Poland	Warsaw Stock Exchange	WIG20	2.27	2.41	2.41	5.94
Portugal	Euronext Lisbon	PSI20	2.05	2.09	1.98	5.72
Romania	Bucharest Stock Exchange	BET	2.03	2.37	2.22	16.61

(continued)

Table 2.1 (continued)

Country	Exchange	Index	Historical VaR (%)	Gaussian VaR (%)	Modified VaR (%)	Range (%)
Russia	Moscow Exchange	MOEX	2.53	3.10	1.73	78.74
Slovakia	Slovakia Stock Exchange	SAX	1.71	1.85	1.62	14.27
Spain	Bolsa de Madrid	IBEX	2.34	2.47	2.24	10.44
Sweden	Stockholm Stock Exchange	OMXS	2.20	2.25	2.10	7.02
Switzerland	Swiss Exchange	SMI	1.70	1.83	1.68	8.72
Turkey	Istanbul Stock Exchange	XU100	2.57	2.62	2.59	2.07
UK	London Stock Exchange, FTSE 100	UKX	1.82	1.95	1.78	9.54
Ukraine	Ukraine Stock Exchange	UX	2.65	3.06	2.65	15.65

Table 2.2 Forecast errors for different VaR estimation approaches in selected European markets

Error metric	Historical VaR (%)	Gaussian VaR (%)	Modified VaR (%)
Absolute error (Delta)	0.79	0.78	0.84
Mean Absolute Percentage Error, MAPE	38.70	35.88	43.27
Root Mean Squared Error, RMSE	0.92	0.90	0.98

As expected, the ES metrics for each market are notably higher than the VaR metrics as they give idea for the riskiest part of the distribution's tail. The stock indices in Ukraine, Russia and Turkey have the highest ETL estimates with 4.57%, 4.56%, and 3.85% historical ES. Overall the expected shortfalls in different markets fall in the range between 2.6% and 4.6% with indices in large and developed markets registering values from 2.9% to 3.4%. Markets in countries like Greece, Italy and Spain are naturally above that. A second insight from the results is that the difference in ES estimates between the three approaches under study are much more pronounced than they are in the VaR estimates. The lowest difference ranges are around 12–17%, while the larges difference can be around or even above 100%. Particularly large discrepancies are observed in the Baltics, Russia, and Bulgaria. The focus on the

Table 2.3 Daily 95% expected shortfall (ES) metrics for selected European stock markets

Country	Exchange	Index	Historical ES (%)	Gaussian ES (%)	Modified ES (%)	Range (%)
Belgium	Euronext Brussels	BEL20	3.15	2.66	3.90	46.67
Bulgaria	Bulgarian Stock Exchange	SOFIX	2.86	2.30	4.47	94.54
Czechia	Prague Stock Exchange	PX	3.31	2.80	1.89	74.93
Estonia	Estonian Stock Exchange	OMXT	2.57	2.11	1.39	84.70
Finland	Helsinki Stock Exchange	HEX	3.24	2.77	3.47	25.25
France	Euronext Paris	CAC40	3.42	2.91	3.23	17.46
Germany	Deustche Boerse	DAX	3.32	2.83	2.99	17.13
Greece	Athens Exchange	ATH	4.80	4.16	5.49	32.05
Hungary	Budapest Stock Exchange	BUX	3.50	3.12	3.45	11.99
Italy	Borsa Italiano	FMIB	3.87	3.32	4.91	48.00
Latvia	Latvian Stock Exchange	OMXR	2.88	2.57	1.43	101.0
Lithuania	Vilnius Stock Exchange	OMXV	2.55	2.06	1.21	111.9
Netherlands	Amsterdam Stock Exchange	AEX	3.21	2.69	3.13	19.44
Norway	Oslo Stock Exchange	OSEAX	3.60	2.94	4.63	57.60
Poland	Warsaw Stock Exchange	WIG20	3.49	3.02	4.24	40.26
Portugal	Euronext Lisbon	PSI20	3.07	2.62	3.41	30.13

(continued)

Table 2.3 (continued)

Country	Exchange	Index	Historical ES (%)	Gaussian ES (%)	Modified ES (%)	Range (%)
Romania	Bucharest Stock Exchange	BET	3.66	2.98	4.12	38.14
Russia	Moscow Exchange	MOEX	4.56	3.90	1.73	163.5
Slovakia	Slovakia Stock Exchange	SAX	2.85	2.32	1.93	47.99
Spain	Bolsa de Madrid	IBEX	3.56	3.10	3.41	14.89
Sweden	Stockholm Stock Exchange	OMXS	3.31	2.83	3.31	17.18
Switzerland	Swiss Exchange	SMI	2.72	2.30	2.80	21.68
Turkey	Istanbul Stock Exchange	XU100	3.85	3.30	4.30	30.15
UK	London Stock Exchange, FTSE 100	UKX	2.91	2.45	2.86	18.69
Ukraine	Ukraine Stock Exchange	UX	4.57	3.85	3.11	47.26

distribution's tail where the analyst usually has much less data to work with makes the choice of estimation approach a significantly more important one.

The three ETL estimation methods can be compared by calculating the error rates between forecasts based on data from 2006 to 2019, and the 2020 baseline realized ETL. Again, the year 2020 is particularly interesting as it presents a significant economic turmoil—exactly the type of environment the expected shortfall metric is conceived for. Forecast errors are calculated and presented in Table 2.4. The non-parametric historical estimate consistently outperforms the other two approaches. Its absolute error, MAPE and RMSE is notably lower than the competing alternatives. Both parametric methods register higher errors and in terms of absolute errors and RMSE are practically indistinguishable from one another.

Thus, in the case of European markets, the analyst may do better by relying purely on historical data than *a priori* imposing assumption on the distribution—regardless of whether those assumptions are of normality or non-normality. Concluding the excursion into VaR and ES estimation methods, one reaches the uncomfortable conclusion that even long time series of data including deep economic recessions

Table 2.4 Forecast errors for different ES estimation approaches in selected European markets

Error metric	Historical ES (%)	Gaussian ES (%)	Modified ES (%)
Absolute error (Delta)	1.32	1.72	1.70
Mean Absolute Percentage Error, MAPE	40.83	62.52	73.51
Root Mean Squared Error, RMSE	1.49	1.87	1.86

underestimated the rise in risk in 2020 by about 40%. Yet again this points to the fact that risk metrics need to be considered more as rough guides than as precise measurements, especially in times of crises.

2.6 Conclusion

This chapter presented a short excursion into the most traditional risk metrics. Their very existence is a reflection of humans' strong desire to understand and tame uncertainty, which is prevalent in all aspects of business and life. Almost all activities contain a component of risk—some event that may or may not occur and all the risk metrics try to capture some crucial aspect of this very occurrence. Using qualitative estimates may be a good starting point but a truly precise and rigorous analysis requires one to leverage more sophisticated quantitative methods. The language of probability is often used to measure risk. This stems from the assumption that uncertainties may be treated as a random variable that can be described through a certain, often unknown, statistical distribution.

The natural starting point for describing risks has been the normal distribution, which is useful for well-defined symmetric events that are subject to physical limits. Using the familiar bell-shaped curve has not proved enough. It posits small probabilities for rare events which can be unrealistic, and many analysts resort to fat-tailed distributions that remedy this weakness. Some real-life phenomena are not really bounded and many processes in business, and particularly digital business, are characterized by a winner-takes-all dynamic, producing highly skewed outcomes. Distributions of the power law type are well-suited for modeling such situations and give the analyst the ability to represent rather well phenomena such as popularity, damage, inequality, and many others.

After conceptualizing risk as a random variable, one may easily calculate metrics that capture its many aspects. Traditionally, volatility metrics such as the variance or the standard deviation have been used as a first pass to estimate average deviation from one's expectations, and indicators such as the beta give an idea of volatility compared to a benchmark. Another strand of measurement has focused on estimating the risk-return tradeoff, and the simplest metric to do this is the Sharpe ratio. Finally, there is a group of metrics that aims to give a probabilistic idea of the maximum expected loss over some time. The Value at Risk and the Expected Shortfall metrics fall under this group. Since the true distribution of the risk under study is not known, there are a number of ways to estimate them, and they yield quantitative results which are sometimes at significant variance with one another. This serves to underline both the important albeit subjective role of the risk management analyst in the process, as well as to show that even metrics that look enticingly accurate are mere approximations to a complex phenomenon under study.

References

1. Abbas, Q., Ayub, U., Saeed, S.K.: CAPM-exclusive problems exclusively dealt. Interdisc. J. Contemp. Res. Bus. **2**(12), 947–960 (2011)
2. Acerbi, C., Tasche, D.: On the coherence of expected shortfall. J. Bank. Finance **26**(7), 1487–1503 (2002)
3. Akter, N., Nobi, A.: Investigation of the financial stability of S&P 500 using realized volatility and stock returns distribution. J. Risk Financ. Manage. **11**(2), 22 (2018)
4. Bebchuk, L.A., Cohen, A., Spamann, H.: The wages of failure: executive compensation at Bear Stearns and Lehman 2000–2008. Yale J. Reg. **27**, 257 (2010)
5. Berghel, H.: Equifax and the latest round of identity theft roulette. Computer **50**(12), 72–76 (2017)
6. Bouchaud, J.P.: Power laws in economics and finance: some ideas from physics (2001)
7. Chandola, V., Banerjee, A., Kumar, V.: Anomaly detection: a survey. ACM Comput. Surv. (CSUR) **41**(3), 15 (2009)
8. Chang, C.L., Jiménez-Martín, J.Á., Maasoumi, E., McAleer, M., Pérez-Amaral, T.: Choosing expected shortfall over VaR in Basel III using stochastic dominance. Int. Rev. Econ. Financ. **60**, 95–113 (2019)
9. Chen, J.: On exactitude in financial regulation: Value-at-Risk, expected shortfall, and expectiles. Risks **6**(2), 61 (2018)
10. Chernobai, A., Rachev, S.T., Fabozzi, F.J.: Operational risk. In: Encyclopedia of Financial Models (2012)
11. Child, D.: The Essentials of Factor Analysis. A&C Black, USA (2006)
12. Crouhy, M., Galai, D., & Mark, R. (2006). *The essentials of risk management* (Vol. 1). New York: McGraw-Hill.
13. Daníelsson, J., Jorgensen, B.N., Samorodnitsky, G., Sarma, M., de Vries, C.G.: Fat tails, VaR and subadditivity. J. Econometr. **172**(2), 283–291 (2013)
14. Erfani, S.M., Rajasegarar, S., Karunasekera, S., Leckie, C.: High-dimensional and large-scale anomaly detection using a linear one-class SVM with deep learning. Pattern Recogn. **58**, 121–134 (2016)
15. Elton, E.J., Gruber, M.J., Brown, S.J., Goetzmann, W.N.: Modern Portfolio Theory and Investment Analysis. Wiley (2009)
16. Eom, C., Kaizoji, T., Scalas, E.: Fat tails in financial return distributions revisited: evidence from the Korean stock market. Phys. A **526**, 121055 (2019)
17. Esterhuysen, J.N., Styger, P., Van Vuuren, G.: Calculating operational value-at-risk (OpVaR) in a retail bank. S. Afr. J. Econ. Manage. Sci. **11**(1), 1–16 (2008)
18. Fabozzi, F.J., Gupta, F., Markowitz, H.M.: The legacy of modern portfolio theory. J. Invest. **11**(3), 7–22 (2002)
19. Fama, E.F., French, K.R.: The capital asset pricing model: theory and evidence. J. Econ. Perspect. **18**(3), 25–46 (2004)
20. Farinelli, S., Ferreira, M., Rossello, D., Thoeny, M., Tibiletti, L.: Beyond Sharpe ratio: optimal asset allocation using different performance ratios. J. Bank. Finance **32**(10), 2057–2063 (2008)

21. Farmer, J.D., Geanakoplos, J.: Power laws in economics and elsewhere. In: Santa Fe Institute, May (2008)
22. Fernandez, P.: CAPM: an absurd model. Bus. Valuat. Rev. **34**(1), 4–23 (2015)
23. Gabaix, X.: Power laws in economics and finance. Annu. Rev. Econ. **1**(1), 255–294 (2009)
24. Gabaix, X.: Power laws in economics: an introduction. J. Econ. Perspect. **30**(1), 185–206 (2016)
25. Gogoi, P., Bhattacharyya, D.K., Borah, B., Kalita, J.K.: A survey of outlier detection methods in network anomaly identification. Comput. J. **54**(4), 570–588 (2011)
26. Gualandi, S., Toscani, G.: Human behavior and lognormal distribution. A kinetic description. Math. Models Methods Appl. Sci. **29**(04), 717–753 (2019)
27. Guharay, S., Chang, K.C., Xu, J.: Robust estimation of value-at-risk through distribution-free and parametric approaches using the joint severity and frequency model: applications in financial, actuarial, and natural calamities domains. Risks **5**(3), 41 (2017)
28. Habeeb, R.A.A., Nasaruddin, F., Gani, A., Hashem, I.A.T., Ahmed, E., Imran, M.: Real-time big data processing for anomaly detection: a Survey. Int. J. Inf. Manage. **46**, 289–307 (2018)
29. Hansson, S.O.: Risk: objective or subjective, facts or values. J. Risk Res. **13**(2), 231–238 (2010)
30. Huang, J., Ding, A., Li, Y., Lu, D.: Increasing the risk management effectiveness from higher accuracy: a novel non-parametric method. Pacific-Basin Finance J. 101373 (2020)
31. Huisman, R., Koedijk, K.G., Pownall, R.: VaR-x: fat tails in financial risk management. J. Risk **1**(1), 47–61 (1998)
32. Johnson, S.L., Faraj, S., Kudaravalli, S.: Emergence of power laws in online communities. MIS Q. **38**(3), 795-A13 (2014)
33. Jorion, P.: Value at Risk: The New Benchmark for Managing Financial Risk, vol. 2, 3rd edn. McGraw-Hill, New York (2006)
34. Koike, T., Hofert, M.: Markov chain Monte Carlo methods for estimating systemic risk allocations. Risks **8**(1), 6 (2020)
35. Kokoris, A., Archontakis, F., Grose, C.: Market risk assessment. J. Risk Finance Incorporat. Balance Sheet **21**(2), 111–126 (2020)
36. Leone, P., Porretta, P., Vellella, M. (eds.): Measuring and Managing Operational Risk: An Integrated Approach. Springer (2018)
37. Ling, X.: Normality of stock returns with event time clocks. Account. Finance **57**, 277–298 (2017)
38. Liu, Z., Moghaddam, M.D., Serota, R.A.: Distributions of historic market data–stock returns. Eur. Phys. J. B **92**(3), 1–10 (2019)
39. Markowitz, H.: Portfolio selection. J. Finance **7**, 77–91 (1952)
40. McGoun, E.G.: The history of risk "measurement." Crit. Perspect. Account. **6**(6), 511–532 (1995)
41. Mishra, S., Chawla, M.: A comparative study of local outlier factor algorithms for outliers detection in data streams. In: Emerging Technologies in Data Mining and Information Security, pp. 347–356. Springer, Singapore (2019)
42. O'Sullivan, P.: The capital asset pricing model and the efficient markets hypothesis: the compelling fairy tale of contemporary financial economics. Int. J. Polit. Econ. **47**(3–4), 225–252 (2018)
43. Omar, S., Ngadi, A., Jebur, H.H. (2013). Machine learning techniques for anomaly detection: an overview. Int. J. Comput. Appl. **79**(2)
44. Peiró, A.: The distribution of stock returns: international evidence. Appl. Financ. Econ. **4**(6), 431–439 (1994)
45. Peterson, B.G., Carl, P., Boudt, K., Bennett, R., Ulrich, J., Zivot, E., Cornilly, D., Hung, E., Lestel, M., Balkissoon, K., Wuertz, D.: *Package "Performance Analytics"*. R Team Cooperation (2018)
46. Perold, A.F.: The capital asset pricing model. J. Econ. Perspect. **18**(3), 3–24 (2004)
47. Phua, C., Lee, V., Smith, K., Gayler, R.: A comprehensive survey of data mining-based fraud detection research (2010). arXiv:1009.6119
48. Qiu, J., Wu, Q., Ding, G., Xu, Y., Feng, S.: A survey of machine learning for big data processing. EURASIP J. Adv. Signal Process. **2016**(1), 67 (2016)

49. Rahl, S.: Value-at-risk: a dissenting opinion. In: Rahl (ed.) Risk Budgeting: Risk Appetite and Governance in the Wake of the Financial Crisis. Incisive Media (2012)
50. Rousseeuw, P.J., Hubert, M.: Anomaly detection by robust statistics. Wiley Interdisc. Rev. Data Min. Knowl. Discov. **8**(2), e1236 (2018)
51. Rowe, D.: The false promise of expected shortfall. Risk **25**(11), 58 (2012)
52. Rubinstein, M.: Markowitz's "portfolio selection": a fifty-year retrospective. J. Financ. **57**(3), 1041–1045 (2002)
53. Safi, R.: The Equifax Data Breach: A Corporate Social Responsibility Perspective and Insights from Tweets for Cybersecurity Incident Management (2020)
54. Schubert, E., Sander, J., Ester, M., Kriegel, H.P., Xu, X.: DBSCAN revisited, revisited: why and how you should (still) use DBSCAN. ACM Trans. Database Syst. (TODS) **42**(3), 1–21 (2017)
55. Silver, N.: Blindness to risk: why institutional investors ignore the risk of stranded assets. J. Sustain. Finance Investment **7**(1), 99–113 (2017)
56. Sollis, R.: Value at risk: a critical overview. J. Financ. Regulat. Compliance **17**(4), 398–414 (2009)
57. Thaler, R.H., Sunstein, C.R.: Nudge: Improving Decisions About Health, Wealth, and Happiness. Penguin (2009)
58. Thurner, S., Szell, M., Sinatra, R.: Emergence of good conduct, scaling and Zipf laws in human behavioral sequences in an online world. PLoS ONE **7**(1), e29796 (2012)
59. Verhoeven, P., McAleer, M.: Fat tails and asymmetry in financial volatility models. Math. Comput. Simul. **64**(3–4), 351–361 (2004)
60. Wang, R., Zitikis, R.: An axiomatic foundation for the expected shortfall. Manage. Sci. (2020)
61. West, G.B.: Scale: The Universal Laws of Growth, Innovation, Sustainability, and the Pace of Life in Organisms, Cities, Economies, and Companies. Penguin (2017)
62. Yamai, Y., Yoshiba, T.: On the validity of value-at-risk: comparative analyses with expected shortfall. Monetary Econ. Stud. **20**(1), 57–85 (2002)

Chapter 3
Risk in Digital Assets

3.1 Introduction

The digital transformation of modern economies continues apace. By 2023 as much as 65% of global GDP may be digitized, and the trend is expected to robustly continue afterwards.[1] This is often touted as a key boost to productivity and hence economic well-being. On the flip side the digital economy engenders many new interactions, intensifies information flows, and presents one with unique digital assets. All of these serve to greatly increase the risk exposure of organizations and individuals by both modifying probability of occurrence, and amplifying impact. Sometimes this is half-jokingly referred to as the situation in which winners win exponentially, but losers lose exponentially. On the analytic front, this puts new challenges and requires new approaches and tools for modeling digital risks. The natural starting point in this discussion would be the realization that in the digital world phenomena virtually do not tend to be physically bounded and can thus have unpredictable dynamics. Most notably, winner-takes-all phenomena, characterized by power laws or similar distributions are expected to be prevalent—ranging from the number of visits per website, sharing of online information, through revenue of digital companies, all the way into expected losses from data breaches (see e.g. Fig. 3.1).

Even mundane variables that one does not expect to be exponentially distributed can be. A case in point is the total amount spent in online retail shopping. While this should be bounded by the ability of users to consume, in fact the amounts vary enormously, with very few customers with exorbitant spending dominating the tail of the distribution. This alerts the analyst to the fact that many conventional modeling assumptions may fail to hold in the digital world.

One may conceptually divide the many cyber-risk as falling into two large groups. The first one concerns risks that materialize in real life but are engendered, amplified, or enabled, in the digital realm. Among the key examples here are exerting influence on online users for advertising or political purposes and leveraging the results of

[1] Fitzgerald et al. [12].

A. Gerunov, *Risk Analysis for the Digital Age*, Studies in Systems, Decision and Control 219, https://doi.org/10.1007/978-3-031-18100-9_3

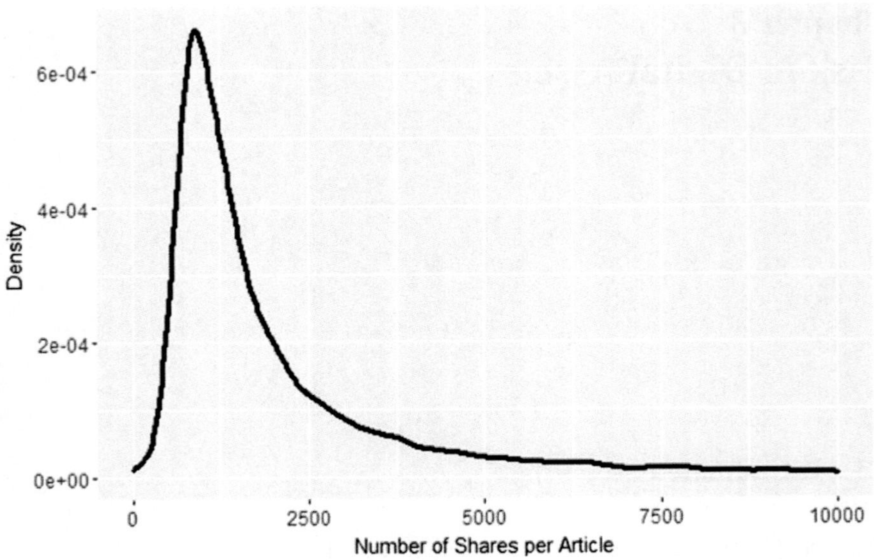

Fig. 3.1 Distribution of shares per online article in the Fernandes et al. [11] Data

automated decision-making. In those instances, an IT solution creates a situation of risk (e.g. possibility of discrimination based on automated credit scoring) but the results affects individuals (e.g. lost opportunity) and organizations (e.g. foregone profits). Those risks are better dealt with by analyzing real-world consequences and estimating probabilities and impacts using traditional approaches. The other large set of risks occur primarily in the digital environment and can be fully analyzed within its reach without taking recourse to physical real-world phenomena. This does not come to say that these do not exist; it merely asserts that their physical world implications are not strictly necessary for the practical treatment of risks. For instance, digital assets—both non-exchange traded (like databases and algorithmic models) as well as traded (like tokens and cryptocurrencies) are good cases in point. While the dynamics of crypto exchanges clearly has significant real life repercussions, the analyst may model them rather fully by taking into account observed returns distributions and related characteristics. The set of risks stemming from digital assets that are only loosely bounded by physical reality is the challenge in modeling. Oftentimes, traditional approaches may not be sufficient, and the modeler will either have to supplement them or do a full-fledged assumption-laden estimation in order to get an understanding of their dynamics.

This chapter thus reviews the different types of information and digital assets that engender new and challenging types of risks for organizations and individuals. Their origin and drivers are outlined in an attempt to aid further modeling. The following sections address two major questions or risk analysis—how to valuate digital assets with no market for the purposes of risk management, and how to assess emerging assets with limited market information and significant volatility.

3.2 Information Assets

The process of digital transformation can be loosely conceptualized as the organization's effort to structure its business processes around its digital information assets. Those information assets consist of all the data, models, information, processing capacity and IT infrastructure and in the modern digital economy may form the basis of competitive advantage in business. Their benefits need to be contrasted to the ensuing costs and risks that stem from them, which results in a balanced decision whether or not to retain a specific information asset. An apt case in point are data assets in European companies. Under the European Union's comprehensive privacy legislation, the General Data Protection Regulation (GDPR), companies are tasked to process only strictly necessary personal data. While organizations may feel the temptation to hoard excessive amounts of data in the hope that this will eventually unlock value, legislation tasks them to only use what is required to generate a specified benefit. Thus, decreasing the data stock leads to a decrease in data governance and curation costs in addition to improved compliance. On the other hand, it is uncertain whether a long time series of data may be needed for some unforeseen future application, which makes it problematic to dispose of data indiscriminately. Such crucial tradeoffs in situations of uncertainty fall squarely in the focus of modern risk management for the digital age.

3.2.1 Types of Information Assets

The natural starting point of analyzing information asset is to appreciate their diversity, scope of application, and utility. Those assets can be divided into four majors groups, as follows[2]:

- **Data assets**—those include databases, data and system documents, instructions and training materials, standard operating procedures, operational and enterprise plans and programs, as well as printed matter and backed-up data and media. Those data assets are key for the functioning of the modern organization as they can represent unique and difficult to imitate resources that competitors do not have access to. All manners of organizations collect and maintain (sometimes immense) databases of their operations, transactions, and customers that are used both to manage and improve activities, as well as to feed into strategic and tactical decision-making. Information on the internal workings and individuals clients is of immense value and may turn to be a key competitive advantage, e.g. through the practice of product personalization. With the advent of machine learning and automated analytics, data assets gain even more prominence. They can be used to feed into statistical algorithms in order to train models capable of executing domain-specific forecasting and classification tasks with high performance. In

[2] Shamala and Ahmad [33], Chen et al. [8].

this manner, the trained models may be able to offer correct forecast of fraud and default rates, demand levels, or purchase propensity, to name a few examples. The destruction of data puts those activities at risk, while the leakage of data makes them easily imitable, thus eating away at the strategic advantage of possessing them. The risks are even exacerbated as one moves from the individual or organizational level onto the state level where sensitive personal data may be abused by adversaries in order to gain economic, political, or military advantage. In short, among all information assets, data assets are by far the most valuable and the risks stemming from their compromise are likely the ones with the highest impact.

- **Software assets**—this group consists of applications, system software, development tools, libraries, packages, and open-source software. Software assets can be broadly divided into custom software and commercial off-the-shelf software (COTS). The former is a proprietary asset to the organization and may even comprise intellectual property. Insofar as it embodies key knowledge and competences and is a product of a lengthy and expensive development process, custom software is a key assets and risks associated with it may have critical repercussions. Moreover, if the codebase of the proprietary programs is also the foundation of a digital product, its compromise may affect not only the owner, but also their partners, customers, and vendors. The result is a spillover of risk over to a network of stakeholders with a potentially large impact. Moreover, the spillover obliterates independence of events and can create a domino effect. The latter group of assets—COTS—consists of programs purchased on the open markets. While they may be compromised themselves, the organization may simply choose to patch, update or replace them. As a result, operational costs are incurred but little long-term effects on strategic positioning should ensue. The main risk in these cases consists of data compromise with potentially deleterious consequences, as outlined above.
- **Physical assets**—this group roughly corresponds to the organizational IT infrastructure and includes computing and processing equipment (servers, mainframes, end user devices), communication equipment, storage media, peripherals, and other equipment such as data center equipment and related power generation. Physical assets are usually almost perfectly replicable by easy market procurement and are thus not crucial for competitive positioning. The main risks they pose are possible outages and downtimes that may affect operations. This is likely to affect productivity for a limited time interval, but the overall expected effects of those uncertain negative outcomes materializing are usually dwarfed by data- and information-related risk. While the analyst should not make absolute a priori assumptions on risk magnitudes, it is often the case that physical assets carry a well-known probability of malfunction, and a correspondingly limited overall impact.
- **Service assets**—this set includes networks, communication services, and public facilities such as heating, lighting and electric power. This group of assets is hardly responsible for attaining competitive advantage and is rarely considered crucial in carrying out operations and business processes. It merely enables the smooth functioning of other parts of the digital infrastructure. However, issues

with connectivity or electricity may render the organization paralyzed. Considering this, most crucial service assets offer high availability and backup option so that business continuity is ensured. Only in very extreme circumstances (e.g. catastrophic weather events) are service assets rendered unusable and operations are affected. While this needs to be considered, it is rarely the focus of a digital risk management program.

The understanding and prioritization of information assets when analyzing cyber risks is crucial for obtaining a fuller picture of the overall risk landscape individuals, organizations, and even state actors face in the digital realm. The highest level of risk is likely found with data assets and proprietary software, while generic hardware and configurations seem to be of less importance.[3]

3.2.2 Sources and Types of Risks

The precise identification of risks that endanger information assets is not a straightforward process and may have different starting points—from investigating and evaluating activities, through analysis of vulnerabilities and threats, and all the way into considering the mismatch between security requirements and reality.[4] At any rate, it is widely agreed that the risks stemming from the existence and operation of information assets is mediated by all the four key organizational dimensions: assets themselves, people that relate to them, the processes in place, and the utilized technology.[5] A possible way to identify risks connected to information assets is to analyze the sources of risk, determine if a vulnerability exists and then decide whether there is a credible threat that can exploit that said vulnerability. A sample tree of risk sources is presented in Fig. 3.2. While it may seem that those risks may be typical of any production process, one should note that in a digital environment the threats and assets are virtual, which greatly accelerates the speed of occurrence and mostly removes physical constraints. Moreover, human-related risks dwarf physical and environmental ones.

The first large sub-division is into risks that are predicated on human action or inaction, and those that are not.

- **Human-related risks** are engendered by the willful action of human actors that may be internal or external to the entity in possession of the information assets. It is widely agreed that this group of risks is by far the most prevalent[6] and risk analysis and management for information assets necessarily focuses mostly on people and much less on the underlying technologies, thus coming in stark contrast with what movies and television shows lead the general public to believe.

---

[3] Borek et al. [6].

[4] Cram et al. [10], Joshi and Singh [22], Ali et al. [2], Giat and Dreyfuss [13].

[5] Jaiswal [21].

[6] See Parsons et al. [27] and references therein, as well as Posey et al. [29].

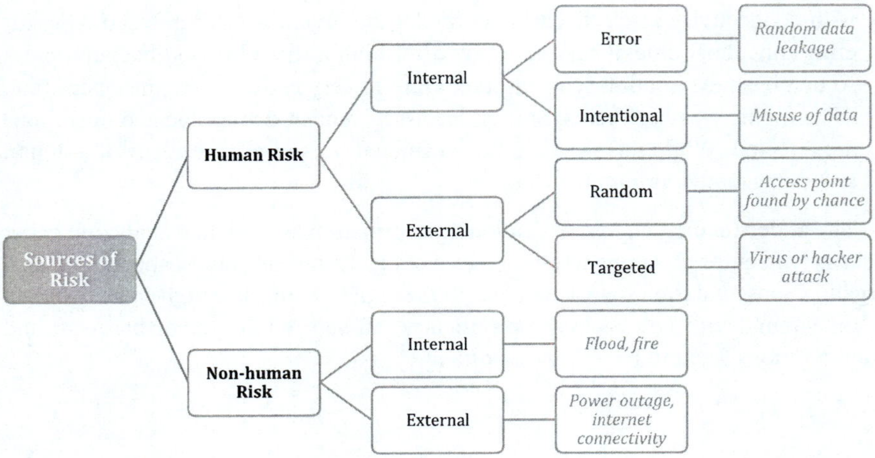

Fig. 3.2 Sources of risk and examples of resulting uncertain outcomes

Some of those risks are internal to the organization. Sometimes human error may compromise information assets and lead to inadvertent effects such as random data leakage, failure of access, or discontinuation of service. Alternatively, malicious insiders, which present the largest threat, may misuse, compromise or even try to sell data assets they have access to. External threats are somewhat rarer but also need to be taken into account. Random ones, for example, can lead to accidental unauthorized access or modification. Those are usually dwarfed by targeted attacks on the outside such as the introduction of malicious software or attempts to gain access to confidential data and exfiltrate it. The latter is a particular risk for large enterprises and governments as they tend to be of interest to well-organized and funded groups of malicious actor (i.e. the so-called advanced persistent threats, APTs).

- **Non-human risk**—the other large set of risks to the information assets comprise of risks that are not directly engendered by human action within the digital ecosystem. Among the internal ones, events such as floods or fires build up this group. External risks here include events such as problems with internet or network connectivity, and power outages. In a well-structured technological architecture, those events should not compromise information assets for a long period of time and may merely make access impossible until addressed. In this case, the analyst would expect a temporary impact with a well-defined probability due to the availability of a relatively long time series of such events.

The key insights when evaluating risk to information assets is that the existence of a vulnerability within the IT ecosystem may not per se constitute a risk.[7] The vulnerability must be exploitable in order to present a potential risk. On the other hand, there must also be a clearly identified threat (e.g. APTs) that could be interested in

[7] Agrawal [1], also, see NIST recommendations in Stoneburner et al. [35].

and capable of exploiting the vulnerability. It is only at this point that the information assets could be considered under risk. Finally, an economic determination needs to be made to explicate motives. On the side of the attacker, the cost of performing the attack must be smaller than the expected probability-adjusted gains. The defenders need only defend if the expected losses are larger than the required mitigation costs. All in all, the analysis of risks to information assets needs to only focus on those that are of sufficient importance to be both attacked and properly protected. For the rest of the ecosystem of potential threats in this complex environment, individuals and organization must learn to accept low impact risks as part of ordinary operations.

3.2.3 Estimating Risk Probability and Impact

While challenging, the identification of risks in information assets is just the usual first step in the risk management process, that is naturally followed by risk evaluation and devising risk mitigation strategies. Qualitative risk evaluation has dominated the narrative about risk as pertaining to information assets.[8] This approach by and large mirrors the approach outlined in Chap. 1 and starts by eliciting ranks for risk probability and impact, with these rankings most often coming from seasoned subject matter experts. The ranking may then be standardized and used within a more quantitative framework for the purpose of estimating the return on security investment (ROSI).[9] These data, however, remain qualitative and imprecise at heart, and is thus very difficult to compare risks with the same rank—i.e. it is unclear if two "high impact" risks carry the same exposure in monetary terms. The preponderance of qualitative evaluations is largely due to reasons of both tradition and convenience. Those estimates have been widely used, and are easy to generate and communicate across the organization. However, their imprecision may engender a false sense of assurance and do more harm than good.[10]

The alternative estimation route for the analyst is to try to get hard data for historically realized probabilities and impacts and then leverage the evaluation tools and metrics presented in Chap. 2. The major challenge here would be the insufficient or even non-existent organizational data for event occurrence which makes it impossible to calculate empirical probabilities. This obstacle can be circumvented by using the large portion of publicly available data for events that lead to a compromise of information assets (e.g. data breaches). Using such datasets the analyst may model the probability distribution of certain events for the general population and then use this as a proxy for the probabilities in the situation under study. This enables the

---

[8] Herrmann [16], Cram et al. [10], Joshi and Singh [22], Allodi and Massacci [3], Tsaregorodtsev et al. [38].

[9] As in Herrmann 16.

[10] Hubbard [18].

analyst to reconstruct a rather accurate approximation of risk probability and feed it into the analytic process.[11]

While it is true that each organization and situation of uncertainty will have their own peculiar characteristics, the probability distribution for the general population is still a crucial signpost that can be adjusted if needed. The risk's impact, however, cannot be easily approximated by looking at other examples. This is due to a simple reason—the vulnerabilities stem from identical technologies and similar processes, while the threat landscape is uniform for many organizations in a given segment. This translates into very similar probabilities of risk occurrence. However, the assets exposed may be extremely different and vary from proprietary software, through deeply trained machine learning models through unique datasets. All of this translates into the fact that the risk impact will be very different and needs to be treated separately as the proportion of asset value lost or the total costs of recovery. This begs the question of what exactly asset values in information assets is and how exactly it can be calculated. The next section briefly addresses those questions.

3.3 Valuation of Information Assets

The digital age has grown tired of the oft-repeated adage that data is the new oil. It is probably meant to underline the importance of using it to feed the creation and personalization of the new digital products and services but at this point it surely strikes the reader as an oversimplification. While data clearly has economic value and needs to be treated as an asset, it is sometimes not exactly clear what this value is.[12] This rings true for many of the information assets reviewed, particularly datasets, models, proprietary software, and digital property rights—they are seldom traded on the markets, and their price can only be approximated after a process of valuation.

It is unclear whether this price is indeed their liquidation price and whether it correctly reflects the value to the organization. This creates a significant challenge for risk analysis—should a negative risk materialize for those assets, rendering them partially or completely unusable, then what is the value of this impact. To answer this question, much attention needs to be devoted to estimating at least an approximate valuation of the information asset at hand—in the presence or absence of market signals.

3.3.1 Market Metrics

As with any assets, the easiest way to price it is to take recourse to the wisdom of the markets. The theory behind this is that the market mechanism has been able to

[11] See e.g. Ikegami and Kikuchi [20].

[12] Wilson and Stenson [40], Peltier [28].

process large amounts of information from both the supply and demand sides and, given the scarcity and utility of the good, has placed an appropriate price. From a practical perspective this is also the easiest and least controversial way to put a valuation on an asset. At least two groups of the information assets are regularly traded on well-established markets—the physical and service assets now have a commodity status and are easily priced by taking reference to the price lists of their many vendors. The case is similar with staple COTS software. However, the assets that are easiest to price are also the least impactful from an organizational perspective. Precisely because they are freely traded, any configuration thereof is very easy to imitate and can hardly constitute a competitive advantage. It is the assets that are rare, valuable, and difficult to imitate that bring the greatest value to the organization and thus present the largest potential impact should a negative event relating to them materialize. Therefore, assets such as proprietary software, IPR, or specific datasets or trained machine learning models are sold less often. Moreover, these assets tend to be unique and thus—more difficult to price appropriately. When having to evaluate such unique assets, the analyst may take some market transactions as benchmarks or indications of the price range. While the benchmark will almost certainly be imprecise it may serve as a useful signpost for valuation. At any rate, should analysts find this benchmark unsatisfactory, they may resort to performing a specific estimation of a given asset's value.

3.3.2 *Expert Estimates*

Whenever there is no clear market where information assets are transacted, or no satisfactory benchmark valuation is readily available, the analyst may resort to valuating the data assets themselves. This process may use one single approach or have a combined (weighted) average of the estimates coming from a number of different approaches. Under perfect information and in an equilibrium markets, all approaches should give the same valuation, reflecting asset fundamentals. In practice, this will not often be the case and the analyst will have to judiciously select which approaches to use, and among those—which ones to favor in the final determination. The possible routes to valuation are as follows[13]:

- **Estimate the cost of producing the information asset**—all the types of information asset have a certain production cost and this may be a suitable way to valuate non-tradable assets. In terms of data assets, the analyst may collate the monetary expense for collecting, curating, processing, and storing data. In the case of proprietary software, risk-adjusted total development costs would be an estimate of the valuation of the asset. Finally, the creation of intellectual property is also related to expenditures and may sometimes appear on the balance sheet of the organization. The cost method is imperfect by definition—after all, the asset

[13] Peltier [28], p. 43.

would be created by a rational actor only if they expect its benefit (hence market value) to outweigh its costs. This means that valuation based on the cost of production will tend to underestimate the true value of the asset. In a traditional setting, the analyst may just add an expected margin of profit to the estimate to reach a more realistic value. In the digital realm this is less straightforward as digital products and revenue based on the information asset may scale exponentially, making it unclear what the appropriate markup is.

- **Calculate the cost of reproducing the asset, if unusable**—in a similar vein to the previous approach, the analyst may decide to estimate the expenses needed for the asset to be reproduced. Unlike the cost of production, which is a backward-looking calculation, the cost of reproduction is a forward-looking one that shows future expected costs. Again, it should contain all the necessary activities that give rise to the asset (e.g. requirements gathering, data collection, development work, business modeling, etc.) that are appropriately and realistically priced. This approach gives the organization an idea of what volume of expenses would be needed to replace the asset. Naturally, the opportunity cost of time needs to be included in this calculation as well—if the organization is unable to reap the benefits of the unusable asset for some time, it foregoes benefits. This foregone profit is an additional cost that needs to be taken into account.
- **Estimate the benefit the information asset brings to the owner**—stemming from the usual approach to asset valuation, an information asset may be valued by the benefits it creates. Just like a physical or financial asset's price is the discounted future net cashflow, the value of data or software may is the discounted net benefit it brings. Most often, this is calculated in monetary terms whereby the cash inflow is the total revenue created by the information asset, and the outflow is the total expenses for the creation, maintenance, and eventual discharge of the asset. Following the accounting logic to its logical conclusion, the analyst may also take into account the amount of depreciation or amortization of the information asset. Just like a physical asset wears and tears over time, so data and software get older and need to be constantly updated. The net benefits approach seems to be the closest approximation to the asset market value and is well-grounded in economic theory. However, this approach may pose a number of practical difficulties and require numerous assumptions to implement (e.g. what part of the organization's profit is derived from this asset). That is why approaches such as estimating the cost of production of the asset tend to be easier to implement—most of the numbers they need are already reported in accounting books and are thus less suspect of subjective and unrealistic assumptions.
- **Estimate the benefit the information asset brings to a competitor or adversary**—an alternative take on the benefits approach is to look not at the owner's benefits but at how useful would this asset be to someone else. This valuation is rather hypothetical and is probably the most heavily-laden with assumptions but gives an indication of whether this asset is of interest to third parties. Should they derive benefit from the asset, then they may be willing to purchase it, thus increasing the price or valuation. If the competitor or adversary has only limited use of the data, model, software, or IPR then this is practically an illiquid asset

with more limited value. The analyst should note that this is a very imprecise approach to working out the valuation as key input parameters are unknown and should rarely (if ever) be used on its own. On the other hand, this may be a suitable starting point for valuation elicitation by experts by nudging them to think in a more global way about valuation. A possible extension here would be to utilize a willingness-to-pay approach whereby the owners of the information asset is asked how much they would be willing to pay to prevent a competitor using a given information against them. Thus the analyst may elicit a concrete number to work with.

- **Work out the cost should the asset become unavailable**—the very existence of the information asset supposes that it brings value to operations and processes. Conversely, if the data, software or hardware configuration become unavailable, this imposes a certain cost on the organization. For example, should the organization's online shopping website become unavailable for four fours, then the cost of that is the profit margin on online sales in this period plus reputational damage. The total costs of asset unavailability thus gives an approximation of its value to the organization and thus—an indication of its valuation. The analyst will again confront the problem of untangling each induvial asset's contribution to the overall value proposition and will have to make a number of assumptions. As with all other approaches, those assumptions need to be clear and explicit so that the valuation may be replicated or reconsidered. At any rate, one many choose to perform a sensitivity analysis on key assumed parameters to see if the valuation remains robust.
- **Calculate the cost of release, alteration or destruction**—a final valuation approach is to calculate the losses that will be incurred in terms of revenue, lost customers, and reputational cost should the risk materialize. This approach is similar to calculating the cost of temporary unavailability but instead focuses on the repercussions of permanent unavailability. Thus, the focus here is not on business process disruption and discontinuity, but rather on the loss of a certain resource. The repercussions may include the loss of certain monetary streams, individual customers or even segments, and reputational and compliance problems. Some simplifying assumptions may have to be made, and the analyst may choose to omit some difficult to calculate intangibles (e.g. reputational loss or employee dissatisfaction) in favor of more tangible costs (e.g. regulatory fine, legal expenses, contractual penalty). While this leads to an underestimate of the asset value, it provides for a less controversial valuation.

All the approaches presented here have a common red thread—the valuation of information assets is a daunting endeavor. If the absence of a vibrant market, the precise value of data, software, hardware, or services may be difficult to contextualize and approximate. At any rate, the analyst must choose an appropriate approach and apply it in a consistent and transparent manner. Cost-based approaches are almost certain to underestimate the value of the asset as the benefits are expected to outweigh the costs. On the other hand, they are easier to apply as they may take advantage of existing accounting data. Benefits-based approaches are the more theoretically sound way to go, but usually entail a somewhat larger set of assumptions. This should not be discouraging—valuation of intangibles, including information assets, is a well-researched topic of practical and academic interest and many established methodologies exist to this end.[14] In any case, a valuation will be more reliable if it is rooted in more than one approach and is able to look at this task from more than one perspective.

3.4 Digital Financial Assets

The accelerating speed of digitization also transforms the way financial value is represented, stored, and traded, giving rise to a whole new class of digital financial assets. While these assets are gaining increasing traction and investor interest, a unified definition and regulation for them is yet to surface.[15] However, all financial digital assets can be loosely defined as sets of binary information that represent financial value.[16] These may be traded on more or less traditional exchanges such as the cryptocurrencies or crypto-securities, or may be exchanged in less traditional ways, which is the case with the non-fungible tokens. Both share two key characteristics: first, a price is usually formed, thus giving an indication of their economic value; second, initial pricing takes place in unique or even one-off situations with possible breaches of rationality. Thus, the market price may not be reflective of fundamentals at all times and bubbles may form.[17]

3.4.1 Types of Digital Financial Assets

Digital financial assets come in many shapes and forms but it may be useful to distinguish between three major types—the group of crypto assets, the electronic money group, and a miscellaneous group (see Fig. 3.3).[18]

The **crypto assets group** consists of digital financial assets that are based upon or enabled by the blockchain technology. In its simplest form, the blockchain is a technology that uses key results from cryptography to create a decentralized digital ledger that can store immutable transactions. Essentially, a block in the chain contains a unique numeric fingerprint (hash) of the whole chain, together with data about the current transaction. Those are then ordered within a larger structure—a chain of blocks (hence the name blockchain). In the plain vanilla version, a number of independent agents hold a version of the blockchain and can verify transactions,

---

[14] For examples, see Wilson and Stenson [40].

[15] Henderson and Raskin [15].

[16] Toygar et al. [36] and references therein.

[17] Gronwald [14].

[18] Milos and Gerasenko [25], Schär [32].

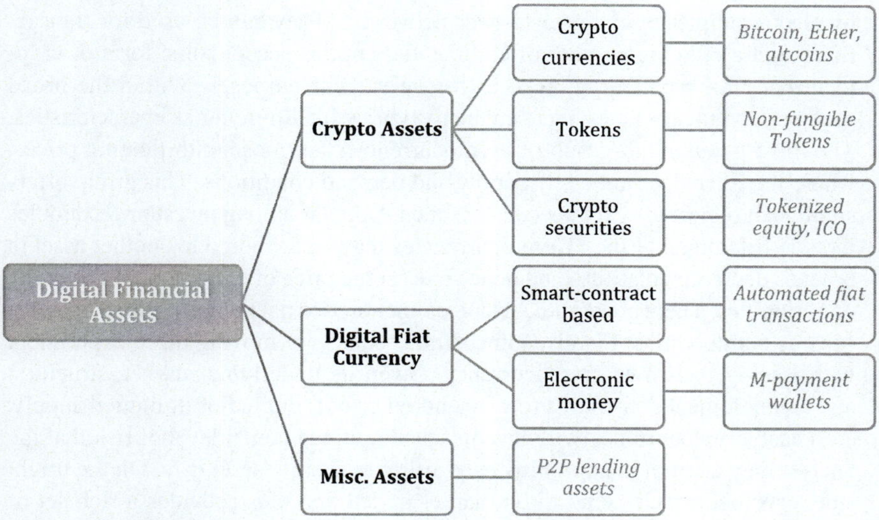

Fig. 3.3 Types of digital financial assets, based on Milos & Gerasenko, [25]; Schär, [32]

including monetary flows and current balances. The first application of the blockchain to create a digital financial asset was the creation of the bitcoin cryptocurrency. In fact, bitcoin became so popular that the general public tends to erroneously use it interchangeably with blockchain, cryptocurrency and crypto assets. Should the public have an unrestricted access to the blockchain, it is called a public blockchain; in the opposite case—it is a private one. Different rules and levels of access to the chain may be imposed—in this case the chain is a permissioned one. All varieties of the blockchain technology are suitable for the creation of specific digital financial assets.

A note of terminology may be worth putting here. In its general sense, a token is a digital representation of an asset, value stream, or a specific activity option. Sometimes the whole crypto assets group is referred to as "tokens" with different uses.[19] To better distinguish between them one could use a set of clearer terms with currency tokens being called cryptocurrencies, security tokens—crypto securities, and utility tokens—just tokens. Each of those digital financial assets may be considered analogous to a real-life analog asset with the cryptocurrencies approximating money, the tokens—approximating intangible assets or options, and the crypto securities—equity.

The crypto assets consist of the following main types of assets:

- **Cryptocurrencies**—those are currencies based on the blockchain technology, whereby the distributed ledger is used to record transactions, calculate balances, execute and settle payments. The cryptocurrencies are usually issued and operated

[19] Conley [9].

by electronic means in a peer-to-peer network.[20] They can be used for transactions, and a growing number of commercial entities accept some form of cryptocurrency for payment, as well as for investment purposes. Within the broad group of cryptocurrencies, there are many types with different risk characteristics. The most popular sub-group of cryptocurrencies are those with floating prices, where the price determined by supply and demand conditions. This group offers significant opportunity for speculation and is popular among investors. Examples include Bitcoin and Ether. Cryptocurrencies may be anchored to another asset in order to decrease volatility (and hence risk) at the price of eliminating large profit opportunities. There are cryptocurrencies anchored to fiat money or to other assets such as commodities. Finally, some central banks are moving on to experiment with issuing their own cryptocurrency. Given its underlying market, structural and technological constraints (e.g. anchored or not, limited or unlimited supply, etc.) each cryptocurrency will have its specific fundamental level of risk that the analyst may attempt to approximate by using relevant market data. On the bright side, cryptocurrencies tend to be market traded and thus provides a rich set of standard financial indicators for them. However, some of the more fringe versions feature very limited trade and using data from such shallow markets often leads to distortions.

- **Tokens**—these are a digital representation of an object or activity of value. Essentially, this is a way to add additional assets to the blockchain (apart from a possible function as a cryptocurrency) and enable the quick transaction and settlement of those assets.[21] A token may represent ownership of a certain work of art, claim to future monetary streams (e.g. profit sharing), or access to a future product or service. Its essential price thus depends on the credibility of the issuer and the promise. To aid trust and decrease risk for the investor, the issuer may choose to take advantage of collateralization. Off-chain collaterals reside outside the blockchain ecosystem (e.g. in an escrow account), while on-chain collaterals reside within the ecosystem, usually binding assets with a smart contract. Should there be no collateral, the token is entirely a promise, and the stemming risk is at its highest. The counterparty risk is not the only major consideration in terms of tokens. It should be noted that while there are some tokens that neatly fall into the pure utility category, this is not always the case. Thus, a utility token may have some features of cryptocurrency or of tokenized equity. Both fall under different sets of strict compliance rules and may thus engender significant regulatory risk. Furthermore, with many tokens it is unclear what regulation they are subject to, which further increases the uncertainty around them. A particular category of tokens is the non-fungible tokens (NFTs) that are used to represent unique assets.[22] In practice this has often turned out to be art or other collectibles. While this is an entirely new market, investors believe that tokenized digital art will appreciate allowing them to profit from their NFTs. As a large part of the value

[20] Ortega [26].

[21] Schär [32].

[22] Trautman [37].

of unique assets is highly subjective, this also spells a correspondingly high level of risk. At any rate, as cryptocurrencies are not crated equal, nor are the crypto tokens. Thus, the analyst must perform modeling on a case-by-case basis, taking into account the fundamentals and specific of each token.

- **Crypto securities**—in their most basic form, tokens are rivalrous digital units of value that entitle their holder to a certain asset or utility. It is only natural that an asset can be equity in a certain company or ownership of a certain security,[23] and so crypto securities are born. By far the most popular form of crypto securities are the different types of tokenized equity. Since they represent company ownership but can still easily be transacted in a decentralized network, those are often seen as a viable way for raising capital for new projects and a possible investment vehicle for organizations and individuals. In a similar way to the initial public offerings (IPOs) of traditional shares, tokenized equity projects have initial coin offerings (ICOs) whereby investors subscribe tokens that may be further traded in a secondary market. Those crypto securities should be subject to standard regulation for securities but some jurisdictions are still struggling with this, thus brining in additional risk. The exact price of those crypto securities is not always clear. From a purely theoretical standpoint it should be equal to the sum of the discounted net cash flows. In practice this is rarely so straightforward—new projects are deviously difficult to valuate precisely, and oftentimes tokenized equity has been fraught possible operational risks or even connected to fraud.[24] In addition to equity, there are also debt instruments and even centralized crypto derivatives. Their valuation is similar to their traditional counterparts[25] but the uncertainty around them is much higher which spells a higher level of expected risk and a correspondingly higher return that is to compensate investors willing to go into the market for crypto assets.

Another large group of digital financial assets is the **digital fiat currency**. It can be defined as an asset stored in electronic form and used as physical currency.[26] While this definition rings a familiar tone with the concept of the cryptocurrency, the digital fiat currency does not necessarily have to be situated on the blockchain and is most often thought of as being issued by a central bank.[27] If this is the case, digital fiat currency may be used as a universal means of settlement across multiple payment networks. This makes it attractive to investors by lowering liquidity risks.

- **Based on smart contracts**—digital fiat currencies may be based on a smart contract, which carries out automated transactions, or even enables the conversion of fiat currency into a digital currency.[28] A smart contract is a specific algorithmic

[23] Roth et al. [31], Liu and Wang [23].

[24] Schär [32].

[25] Catalini and Gans [7].

[26] Bordo and Levin [5].

[27] Raskin and Yermack [30].

[28] Huckle et al. [19].

contract that automatically executes given that some condition is fulfilled.[29] It gained prominence with the proliferation of the blockchain technology and crypto assets, but possible applications go much beyond that. The smart contract consists of code written in a certain programming language and supposes that all parties are in understanding with its content and in agreement with its ramifications. The name, however, is something of a misnomer—the smart contract is neither smart, nor contract. Since it is a coded algorithm, the contract will execute automatically even in cases where both parties agree that execution should not take place. Furthermore, direct observance of stipulated events may not be possible (i.e. they take place outside of the digital realm) and will thus have to be fed into it. Finally, renegotiations that call for an annex in a real-life contract may mean code rewriting in the "smart" one. On the other side, a smart contract cannot yet be enforced in court at the time of writing, making it not exactly a contract. It may be more useful and practical to consider DFCs more as algorithmic than contractual and factor this into the risk analytic exercise. Insofar as one considers central bank-issued DFCs, many of the relevant insights from monetary theory can be applied and a correspondingly lower level of risk could be assumed.

- **Electronic money**—this is a well-known category of digital assets that enjoys both wide use and acceptance. While definitions may vary,[30] e-money is usually considered to be digital money that have a virtual representation, centralized transaction handling and are issued by a monopoly issuer.[31] The issuer may be the central bank or some other certified and regulated entity. Often, electronic money is created on the basis on actual money and is tightly connected to it. A common use case is to store electronic money in a special wallet and carry out transactions through it. Some countries operate central bank-issued electronic money that are only accessible to a narrow range of economic actors such as financial intermediaries, but there is a clear trend to expand the use of central bank-issued e-money to members of the general public. In addition, private providers have made such assets available to consumers in given commercial ecosystems. Generally, money has a lower level of risk as compared to other assets and thus the risk analyst needs to focus its attention on the credibility of the issuer.

Miscellaneous assets—apart from different tokens and versions of electronic money, there is also a plethora of other digital assets that may pique the interest of the modern investor. An example of those would be claims to a cashflow stemming for a centralized or decentralized peer-to-peer (P2P) lending platforms. Those platforms aim to eliminate traditional financial intermediaries such as banks and directly connect lenders with those looking for funds. The loan arrangement has a particular structure given the lending platform and its risk generally stems from both liquidity issues as well as traditional credit risk. It may be impractical to enumerate or even

[29] For more details refer to Hu et al. [17].

[30] Vlasov [39].

[31] Berentsen and Schär [4].

attempt to group all digital financial assets and one who does that is always guaranteed to miss something. However, when the analyst must perform risk modeling on such assets it may be useful to follow a general approach that stands on traditional economic foundations. The key question for any asset is whether it can be used more as a currency or whether it is mostly a security.[32] Should the asset be mostly a currency, then its valuation or price P can be calculated from the familiar quantity equation. If this currency can buy an amount of goods Y, the velocity is denoted as V, and the total amount of currency is M, then the price of one unit is as follows:

$$P = \frac{MV}{Y} \tag{3.1}$$

On the other hand, if the asset mostly resembles a security instrument, whether debt or equity, then its valuation can be estimated through a discounted cash flow (DCF) model. Denoting net cash flows in period i as C_i, and the discount rate as r, and assuming cashflows over n period, then the price P of the digital financial security should be:

$$P = \sum_{i=1}^{n} \frac{C_i}{(1+r)^n} \tag{3.2}$$

As many digital assets are relatively new to the market, the market itself is often shallow, and investors are not always sophisticated, it should come as hardly a surprise that realized prices may be far off their theoretical prediction. The analyst needs to take into account that this novelty serves to increase risk and make asset dynamics much less predictable.

3.4.2 *Types of Risks in Digital Financial Assets*

The novelty of many digital financial assets spells a high level of risk that sometimes may be difficult to estimate given the short time series of their performance. While the types of risk may differ, many of those assets exhibit risks in one or more of the following categories: market risk, technological risk, coordination risk, and extraordinary (or idiosyncratic) risk.[33] Some of those are also common for traditional financial assets while other are specific to digital ones. Furthermore, some of those risks are easier to diversify, while others are harder (see Fig. 3.4).

The group of **market risks** contains risks that tend to affect the overall environment of trading and exchange. Those might be macroeconomic or market-wide phenomena that affect practically all assets that are traded. Examples include the economy going into a recession or the central bank changing the money supply. Those

---

[32] Conley [9].

[33] Shatohina and Kochetkov [34].

Fig. 3.4 Types of risks in digital financial assets, based on Shatohina and Kochetkov [34]

are familiar risks from traditional financial assets but are difficult, if not impossible, to diversify away by investing in alternative digital assets. The most common types of market risks are the following:

- *Price risk*—the uncertainty in the price that the asset will fetch, its dynamics, and the closeness to the actual valuation may engender significant price risk to the investor. Of course, this is a major type of risk that needs to be considered across the board but in the case of digital assets it is somewhat exacerbated by their complexity and novelty.
- *Asset volatility*—valuation uncertainty as well as shifting market expectations for the value of digital assets may lead to significant and unexpected volatility. It can be measured through a given volatility metric (e.g. variance or its square root—the standard deviation) but short time series spell significant uncertainty in the robustness of the estimate.
- *Liquidity risk*—since many markets for digital assets are relatively new or the assets may be unique (e.g. the NFTs) this sometimes translates into insufficient liquidity whereby investors cannot close their positions in a shallow market. This is naturally a major consideration when buying and holding digital assets and means that their balance sheet or mark-to-market pricing may fail to reflect their actual value at liquidation.

Those risks lend themselves to a more quantitative evaluation as market data on pricing, volatility and trade volume exists and the analyst may use it to make inferences.

The loosely labelled group **coordination risks** is also shared with traditional financial assets and pertains to issues of regulation, speculation, and asymmetric information. Since those risks are specific to individual assets (e.g. the regulation is different for security tokens, utility tokens and NFTs), the investor may diversify away those risks by choosing alternative digital assets. The most common types of risks that this group contains are as follows:

- *Regulatory (or Political) risk*—the risk of compliance issues looms large in the case of digital financial assets. Due to high material interest and precedent problematic cases, regulators are increasingly focused on digital financial assets. While there is a clear boundary between currency and securities laws, a given asset may reside in the gray zone, thus engendering risk for its investor. This is much lower for digital fiat currencies and e-money that it is for more complex crypto assets.
- *Speculative risk*—uncertain pricing and high volatility means that investors have a potentially very large upside for entering the digital asset market. Speculation may ensue whereby the price of an asset is artificially inflated or deflated so that it will allow a group of investors to profit as it returns to its baseline. In the crypto world this is tongue-in-cheek called "pump and dump" scheme and may endanger the credulous investor.
- *Informational risk*—this refers to the usual risk of asymmetric information whereby the entity that issues the digital asset or some privileged investors have access to more information than the overall market. This leads to distortions in valuation, pricing, volatility and ensuing losses.

Coordination risks are more difficult to estimate in a quantitative way and the analyst may have to resort to qualitative estimates and expert opinion.

Technological risks are specific to digital financial assets. While traditional securities are now also exclusively digital (virtual), the underlying technology does not underpin their value and dynamics. Conversely, technology underpins the value of digital financial assets and may put constraints on them such as limiting supply (e.g. in Bitcoin) or putting a cap on transaction. Those risks are possibly diversifiable, but this requires a deeper understanding of the IT dimensions of the assets. The major types of technological risks usually fall under one of the following categories:

- *Mining risk*—some crypto assets are created by performing certain tasks in the digital realm, and this creation is called mining. An apt example is the most popular cryptocurrency—Bitcoin, which tasks members of the network to algorithmically solve a mathematical problem and rewards them with units of the currency for its solution. Naturally, the supply of such assets depends on the smooth functioning of the process and gathering a sufficient amount of diverse miners in the network. Failure to do so may lead to skewed outcomes and eventually domination of the network by a few, thus putting all assets within it at risk.
- *Transaction risk*—again, this is typical for digital financial assets where the technology specific lead to different types of transactions. Those may be between anonymous parties, thus cutting the possibility of remediation. Even when this

is not the case, issues of operational security (e.g., recording transactions, transferring keys, protecting against attacks) may create risk when buying or selling those digital assets.

- *Storage risk*—finally, the risk of storing the digital asset is not negligible. The storage itself may be compromised by malicious actors or the owner may lose access to the storage. This is particularly risky with crypto wallets whereby the access keys may not be retrievable, thus spelling the loss of the asset for good.

Technological risks essentially stem from the implementation of the IT solution as well as the actions of its user. Both are subject of extensive study and probabilities for key risks such as failed transactions, number of miners and probability of churn, and storage compromise may be publicly available. Those can be used as benchmarks and approximations for a quantitative or semi-quantitative evaluation of technological risks.

Extraordinary risks are the final group of uncertainties inherent in digital financial assets. Those may be present to a limited extent in classic financial assets but are much more prominent and potentially impactful in the case of digital financial assets, and particularly crypto assets. The group of extraordinary risks includes:

- *Business risk*—this is the risk connected to ordinary business operations that may be disrupted due to a digital financial asset. An apt example are business models that rely on a token for their operation (e.g., claim to rights or purchases) but this token behaves unexpectedly, has limited adoption or transaction problems. All this put the whole business operation in jeopardy, thus creating business risk.
- *Reputational risk*—since digital assets are relatively new and sometimes operate within immature markets fraught with problems and even deceit, dealing in them may lead to reputational risks for the investor. These may be seen as unreliable or excessively speculative, thus tarnishing one's reputation.
- *Fraud risk*—the emergence of innovative financial instruments is not necessarily painless. The initial proliferation of ICOs has attracted numerous malicious actors trying (and often succeeding) to defraud investors.[34] This may be done either through Ponzi schemes, through simply appropriating the proceeds of ICOs or through any other method seen in traditional markets. The added layer of anonymity in digital transactions only exacerbates this possibility and makes it imperative to include fraud risk in the overall analysis.

Those extraordinary risks are clearly diversifiable as one can choose to rely on a stable token in their operations or altogether avoid it, hold only assets from reputable issues and avoid suspect investments. However, effort needs to be expended to manage those risks, and this is made all the more difficult by the fact that usually rigorous quantitative data is not available. The analyst must thus rely on qualitative estimates and allow sufficient buffers for the large uncertainty inherent in them.

[34] Massad [24].

3.4.3 *Estimating Risk Probability and Impact*

Once the analyst has a firm grasp of the risk source and their particularities as it pertains to digital assets, then the next task is to estimate their likelihood and potential impact. In general, there are two distinct groups of cases. Firstly, the digital asset may be a proprietary one with rather limited data availability. The analyst will have to estimate its value and the corresponding loss of value in case of compromise (i.e., risk materialization). A common way to do this is to resort to the methods for valuation of information assets outlined above. On the other hand, the probability of the risk occurrence may be estimated by leveraging internal data (however limited), devising an expert estimate, or using industry or process-level benchmarks. It is often appropriate to combine alternative estimates to gain a better understanding of the actual probability of realization.

In the latter case, the digital asset is a traded one on some exchange. This makes it straightforward to obtain data on its price dynamics, and from there on—reconstruct returns and their statistical distribution. Having this type of information enables the analyst to apply traditional risk metrics and (such as standard deviations, Sharpe ratios, market betas, VaR, ETL, etc.) to the digital assets. However, their novelty and the inability of the market to price them well may skew the information in the sample and result in unreliable metrics that may confuse instead of enlighten, and thus decrease the utility of the risk analysis.

3.5 Risk Modeling for Emerging Assets

The challenges are most visible when performing risk management in high-stakes bets on emerging digital assets such as the cryptocurrencies. These are electronic currencies enabled by the blockchain technology for creating a distributed anonymous ledger that was supposed to enable money without the explicit or implicit guarantee of a central authority (such as governments or Central Banks) and instead rely on the simultaneous transaction validation by the crowd of users. It was only a very short moment of bliss before investors realized that despite their limited value as money, cryptocurrencies are an excellent speculative financial asset. Pandora's box was open. As with many emerging markets, the cryptocurrency one was initially dominated by non-professionals and limited rigorous financial analysis and risk management was in sight. However, the increasing popularity of this type of digital assets engender the need for a formalized risk modeling that is cognizant of the many specifics that cryptocurrencies bear.

3.5.1 Stylized Facts and Traditional Risk Metrics

A natural first step to modeling risk for emerging digital assets is to investigate their dynamics and note the parallels and differences from the dynamics of traditional financial assets. To this end we investigate the twenty largest cryptocurrencies in terms of market capitalization as of 2021 and track their time series back to the fourth quarter of 2013 or when they were introduced. The currencies under scrutiny are Bitcoin Cash, Binance Coin, Bitcoin, Cardano, Dogecoin, EOS, Ethereum Classic, Ethereum, Filecoin, Litecoin, Monero, Neo, Polkadot, Solana, Stellar, Terra, Theta, Tron, Vechain, and Ripple. Their overall dynamics are similar, and it is instructive to have a more detailed study of some of the most popular cryptos.

The story of Bitcoin's price is graphically presented in Fig. 3.5. Bitcoin is the oldest and most popular cryptocurrency and was introduced as early as January 2009. Touted as a new generation of currency for the masses of the future, it underwent a slow start, trading for a few dollars. As late as 2013, the price of Bitcoin below USD 50. As the interest in this class of assets picked up and exchanges opened, this enabled investors not merely to create Bitcoin by mining it, but instead trade it speculatively. This marked an initial spike of interest with the Bitcoin price briefly reaching the USD 20,000 mark in December 2017. The bubble then burst and over 2018 and part of 2019 the price was in free fall, generating significant losses to investors with long positions. Bitcoin then started to recover, and in 2020 it registered significant growth, partly fueled by an influx of amateur investors into the markets. The year 2021 was the cryptocurrency craze that pushed the price of Bitcoin above the USD 60,000 mark before this bubble burst, and the price started to fall. The second panel of Fig. 3.5 shows how trade volume mirrors the price dynamics of bitcoin. As price was low and the digital asset relatively unpopular until 2017, few trades were taking place. Then, investors became cognizant and interested in Bitcoin, and so the trade volume increased exponentially, with every price bubble accompanied by a local peak in trade. As the bubble bursts, so the volume decreases.

Those dynamics are not unique to Bitcoin. Cryptocurrencies in existence likewise lived through a boom-and-bust cycle, reaching a local peak in 2018. This can be clearly seen for Ethereum, Litecoin, Monero, and Ripple in Fig. 3.6. Following a burst of the price bubble in 2018, they also experienced rapid growth in 2020 and 2021 until they reached a peak, followed by a price collapse. Those figures alone indicate to the risk analyst that emergent digital assets are characterized by a very pronounced volatility with significant risks both on the upside and on the downside. This naturally translates into potential for sizeable gains given advantageous timing, but also catastrophic losses. Overall trade volumes of the cryptocurrencies also follow closely the price dynamics.

This visual intuition is also confirmed when looking at the risk and return metrics for the cryptocurrencies under consideration (see Table 3.1). Average daily return for many of them turns negative, and the high volatility is underscored by the large values of the standard deviation. Cryptocurrencies such as Filecoin, Polkadot and Solana all have daily standard deviations of above 20%, and even the least risky assets

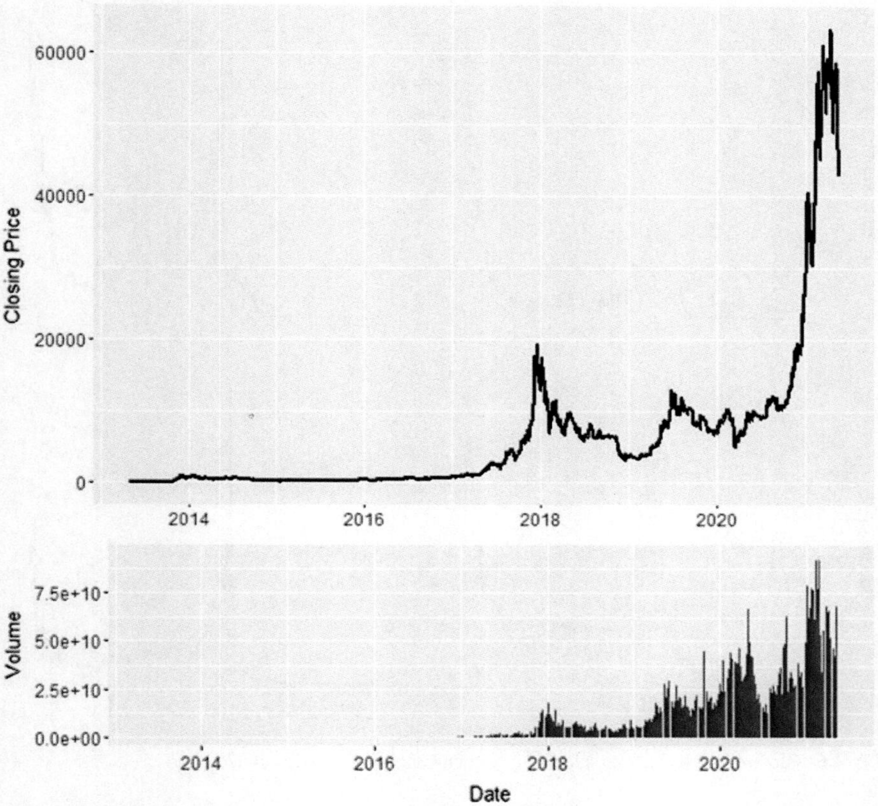

Fig. 3.5 Bitcoin closing prices and trading volume, Q4.2013–Q2.2021

have values above 10%. Ironically, this is Dogecoin—a coin created as a joke by its developers—that turned into one of the most traded cryptocurrencies with a large market capitalization. The statistical properties all the coins under consideration are also showing—both the skewness and the kurtosis point to distributions that are far from the standard Gaussian one.

The standard risk metrics are even more showing. The Value at Risk estimates range from a low of 12% in the case of Ripple to a high of 39% for Polkadot and 41% for Filecoin. Taking into account that this is average daily VaR, such levels of risk may turn out to be overwhelming for many investors. Expected Tail Losses are even higher—the lowest one stands at 15% for Ripple, which the highest goes above 50%. The possibility of losing more than half on one's investment on a bad day is indeed a daunting prospect. Sharpe ratios are probably the easiest way to evaluate the tradeoffs between such high levels of risk and the level of expected return as they show the amount of return per one unit of risk (however estimated). The Sharpe ratios in Table 3.1 use the standard deviation but results are similar for alternative

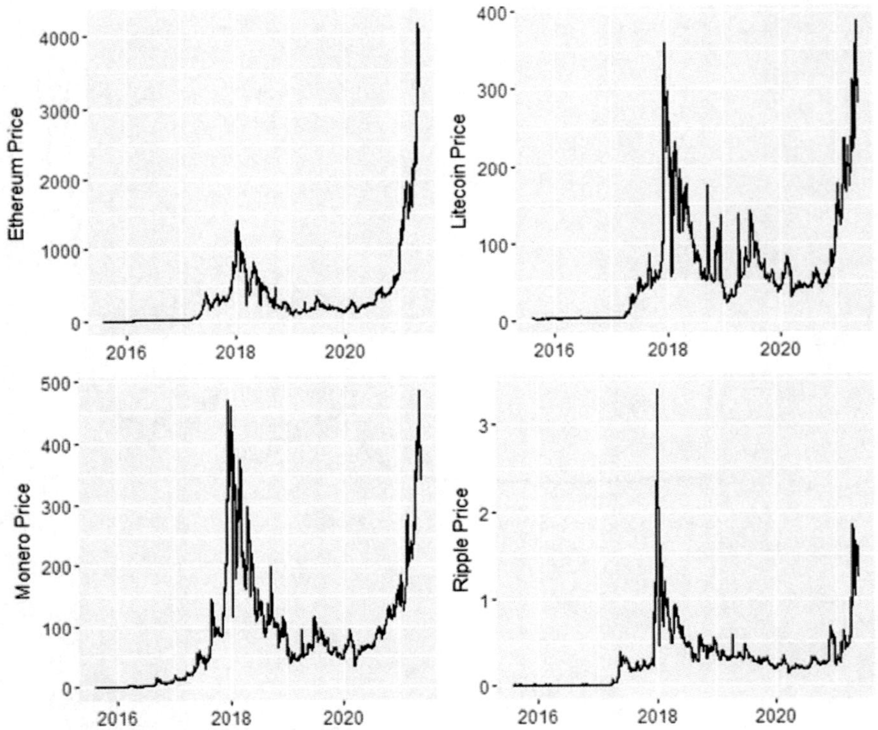

Fig. 3.6 Closing prices of four selected cryptocurrencies, Q4.2015–Q2.2021

denominators, as well. It seems that the coins' tradeoffs are not particularly favorable as many Sharpe ratios are negative, driven by the negative average returns.

In addition, they are relatively small in size with the investors obtaining just 0.02–0.03% point of return more for one percentage point increase in the standard deviation. The avid reader may recall that the average value of the Sharpe ratio for European stock exchanges was more than ten times higher—standing at 0.36. Given objective risk metrics, and the risk-return tradeoffs, it is safe to say that the market for those emerging digital assets is not moved by purely economic rationality but also by animal spirits, ideology, and speculation.

3.5.2 Statistical Properties of the Series

Formal modeling of emerging digital assets should start with eliciting their statistical properties from the data at hand, with the probability distribution being of particular interest. The skewness and kurtosis metrics in Table 3.1 are of significant magnitude and beg the question of whether modeling coins using the normal distribution is an adequate approach. Five different statistical tests for normality are applied to the

Table 3.1 Risk metrics for 20 cryptocurrencies with highest market capitalization

Coin	Mean (%)	Standard deviation (%)	Skewness	Kurtosis	Sharpe	VaR (%)	ETL (%)
Bitcoin cash	−1.44	18.55	−1.330	14.443	−0.078	−31.9	−39.7
Binance coin	−0.59	17.21	−1.795	17.241	−0.034	−28.9	−36.1
Bitcoin	−0.39	15.27	−2.199	25.389	−0.026	−25.5	−31.9
Cardano	0.10	11.94	−0.210	19.060	0.009	−19.5	−24.5
Dogecoin	−0.02	10.69	−0.205	32.136	−0.002	−17.6	−22.1
EOS	−0.57	12.29	−1.727	22.949	−0.046	−20.8	−25.9
Ethereum classic	−0.35	13.17	−1.403	22.708	−0.026	−22.0	−27.5
Ethereum	0.25	16.75	−0.276	16.138	0.015	−27.3	−34.3
Filecoin	−1.95	23.87	−0.488	7.720	−0.082	−41.2	−51.2
Litecoin	−0.40	14.14	−1.594	23.719	−0.029	−23.7	−29.6
Monero	−0.50	13.62	−2.425	21.733	−0.037	−22.9	−28.6
Neo	−0.78	18.28	−1.074	14.705	−0.042	−30.8	−38.5
Polkadot	−0.83	23.25	−0.599	7.712	−0.036	−39.0	−48.7
Solana	1.10	20.55	0.431	7.015	0.054	−32.7	−41.2
Stellar	0.36	10.26	1.775	16.737	0.035	−16.5	−20.8
Terra	0.31	11.34	0.569	9.039	0.027	−18.3	−23.1
Theta	0.44	15.52	0.689	10.246	0.028	−25.1	−31.6
Tron	−0.10	7.76	−0.014	44.939	−0.013	−12.9	−16.1
Vechain	0.31	9.92	−0.071	19.070	0.031	−16.0	−20.1
Ripple	0.12	7.31	1.004	35.248	0.017	−11.9	−15.0

cryptocurrencies under study and results are presented in Table 3.2. Every single test yields highly significant results, rejecting the null hypothesis of normality of the distribution. This points that the appropriate modeling approach, at least for the early stages of emergence of digital assets, cannot take for granted neither the default assumption of gaussian distribution, nor the traditional methods that rely on it. It seems much more appropriate to rely on a skewed distribution with potentially heavy tails that is able to better capture the risk characteristics of the cryptocurrencies under study.

The mutual dynamics of the novel digital coins are also noteworthy. Initially, as the market was trying to price them adequately, the correlations between leading currencies were very high and they were moving almost perfectly together. However, as investor sophistication increased and each coined was evaluated on its own merits, and used as a hedge against others, this knot was untangled and the high mutual correlations all but disappeared.

Table 3.2 Normality test results for 20 selected cryptocurrencies

Coin	Normality tests				
	Anderson Darling	Cramer-von Mises	Pearson	Shapiro Francia	Jarque Bera
Bitcoin cash	530.43	86.36	69,355.38	0.64	490.51
Binance coin	530.53	86.38	69,365.70	0.64	490.52
Bitcoin	1,136.59	245.15	141,166.03	0.00	1.06×10^9
Cardano	535.53	87.71	69,875.18	0.63	490.70
Dogecoin	1,010.13	207.68	120,320.30	0.30	12,197.62
EOS	529.49	86.10	69,259.41	0.64	490.50
Ethereum classic	555.54	93.01	71,912.90	0.62	493.59
Ethereum	662.71	120.64	82,778.61	0.56	596.87
Filecoin	544.26	90.03	70,764.87	0.63	491.52
Litecoin	1,136.59	245.15	141,166.03	0.00	1.06×10^9
Monero	906.43	180.90	108,108.83	0.40	3,198.04
Neo	547.42	90.87	71,086.54	0.63	491.98
Polkadot	984.77	200.91	117,151.54	0.33	8,262.56
Solana	897.96	178.79	107,178.11	0.41	2,931.23
Stellar	858.24	168.97	102,894.15	0.44	2,007.01
Terra	743.48	140.79	90,981.24	0.52	859.99
Theta	549.69	91.46	71,317.31	0.63	492.37
Tron	533.91	87.28	69,710.35	0.63	490.62
Vechain	595.88	103.55	76,008.92	0.60	512.11
Ripple	1,090.65	230.66	131,888.37	0.17	88,315.64

Note All test statistics are significant at levels of $p < 0.0005$

This is quite visible in both the correlation matrix of the four leading coins (Table 3.3) as well as in the correlogram of all twenty (see Fig. 3.7). The rule is that the correlations tend to be of small magnitude and, more often than not, of a negative sign. This points to the fact that while there is some possibility for diversification and hedging within the cryptocurrency market, it is not practically significant in every case. There are obviously some exceptions such as the large negative correlations between Polkadot and Ethereum and Theta, as well as the one between Ripple and Cardano, and a few others. Care must be taken whether these remain stable over time and do not disappear as they start being exploited in the risk management process.

Table 3.3 Correlation matrix of four selected cryptocurrencies

	Bitcoin	Ethereum	Litecoin	Ripple
Bitcoin	1	−0.00688	−0.00628	−0.0151
Ethereum	−0.00688	1	−0.01078	−0.02742
Litecoin	−0.00628	−0.01078	1	−0.00486
Ripple	−0.0151	−0.02742	−0.00486	1

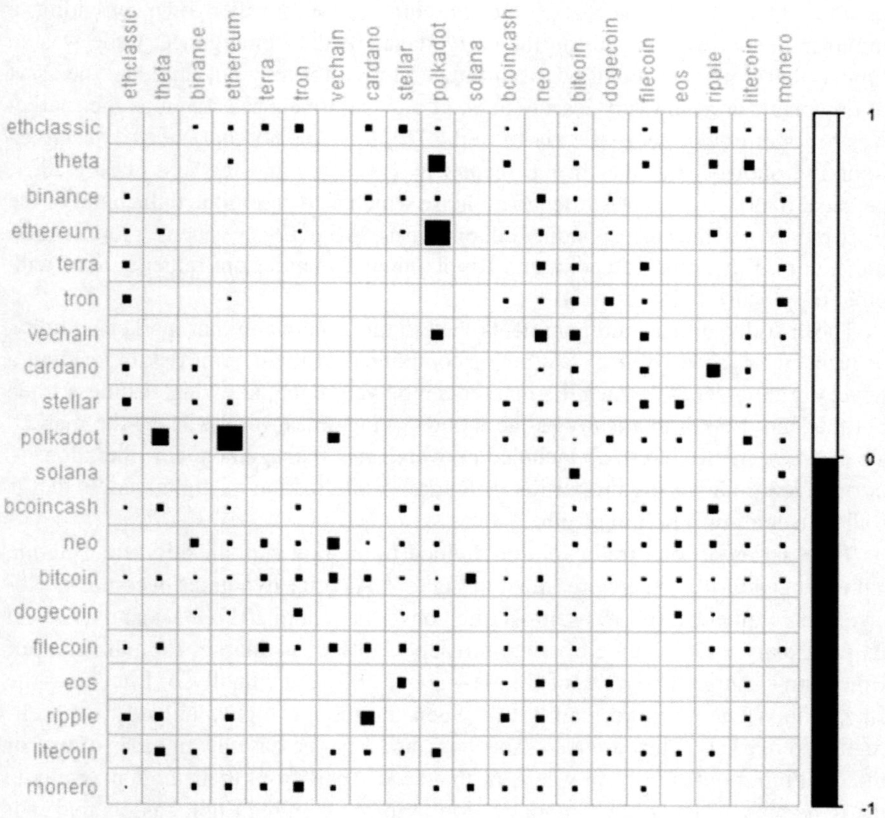

Fig. 3.7 Correlation plot of 20 selected cryptocurrencies

3.5.3 Risk Management Methodology

The stylized facts of the cryptocurrency market reveal that the assets sold there are high risk, extremely volatile, with unstable correlational structure, and non-normal statistical distributions. While the analyst may choose to rely exclusively on traditional risk metrics such as the Sharpe ratio, VaR, or ETL, this may obscure some of the sources of uncertainty. To gain a fuller picture, it is useful to put the coin markets in more context. Traditional financial risk management conveniently rests on the assumption of market liquidity—i.e., that the investor will be able to sell a given asset, taking a hit on the price. However, this may be dramatically different in a market for emerging digital assets. Physical trade volumes for four leading coins are visually presented in Fig. 3.8.

For the initial years of their existence, there was very limited trade, which did not really pick up before 2019. Even as the price bubble for cryptocurrencies was unfolding in late 2017 to early 2018, it was largely driven by relatively small volumes of trade. Only in 2019 and 2020 did the coins came into their own, providing a liquid market. However, during the early stages of their emergence, trade was so limited that it made it a real possibility that an investor may not liquidate the asset *at any price*, or at the very least—do so at a prohibitive price. Even as the market was becoming very deep in early to mid-2020, this was no guarantee of sustained liquidity. Shortly after that in the second half of 2020 and the first one of 2021, physical trade volumes be started to collapse which also precipitated the burst of the price bubble and the corresponding fall of returns. While Fig. 3.8 focuses its attention on Bitcoin, Ethereum, Litecoin, and Ripple, their dynamics are rather similar with other cryptocurrencies, as well.

Those stylized facts and arguments lead to the conclusion that market liquidity, as proxied by trade volume, is a major component of the overall risk of holding a novel digital asset. At points the market can be very deep, allowing seamless trade but this can drop dramatically as the moods swing and investors leave the market. Thus, a new metric has to be included in overall risk management considerations—normalized trade, or the proportion of market capitalization of digital assets that is traded over a given period of time.

Time series of total trade volume divided by market capitalization for Bitcoin, Ethereum and Litecoin are presented in Fig. 3.9. As price dynamics becomes explosive, the volume/cap follows suit—a trend obvious in both 2017–2018 and 2020. At its peak daily trade volume of Bitcoin reaches 80% of its market capitalization, of Ethereum—more than 180%, of Ripple—above 200%, and finally, of Litecoin—the staggering value of more than 300%. Such figures are highly unlikely for well-regulated markets. They are also short lived and can decrease by an order of two or three within months or even weeks of the peak. Apart from their large magnitude, the dynamics of the volume/capitalization ratio of emerging digital assets makes it imperative to include it in overall risk analysis.

The simplest way to go about this is to leverage a volume-adjusted Shape ratio—it is the standard Sharpe that indicates the mean–variance tradeoff but adjusted for the volume/cap ratio. As usually denoting return as r, standard deviation as σ, trade volume as V and market capitalization as C, then the adjusted Shape, S_A, can be expressed as follows:

$$S_A = \frac{r}{\sigma} * \frac{V}{C} \tag{3.3}$$

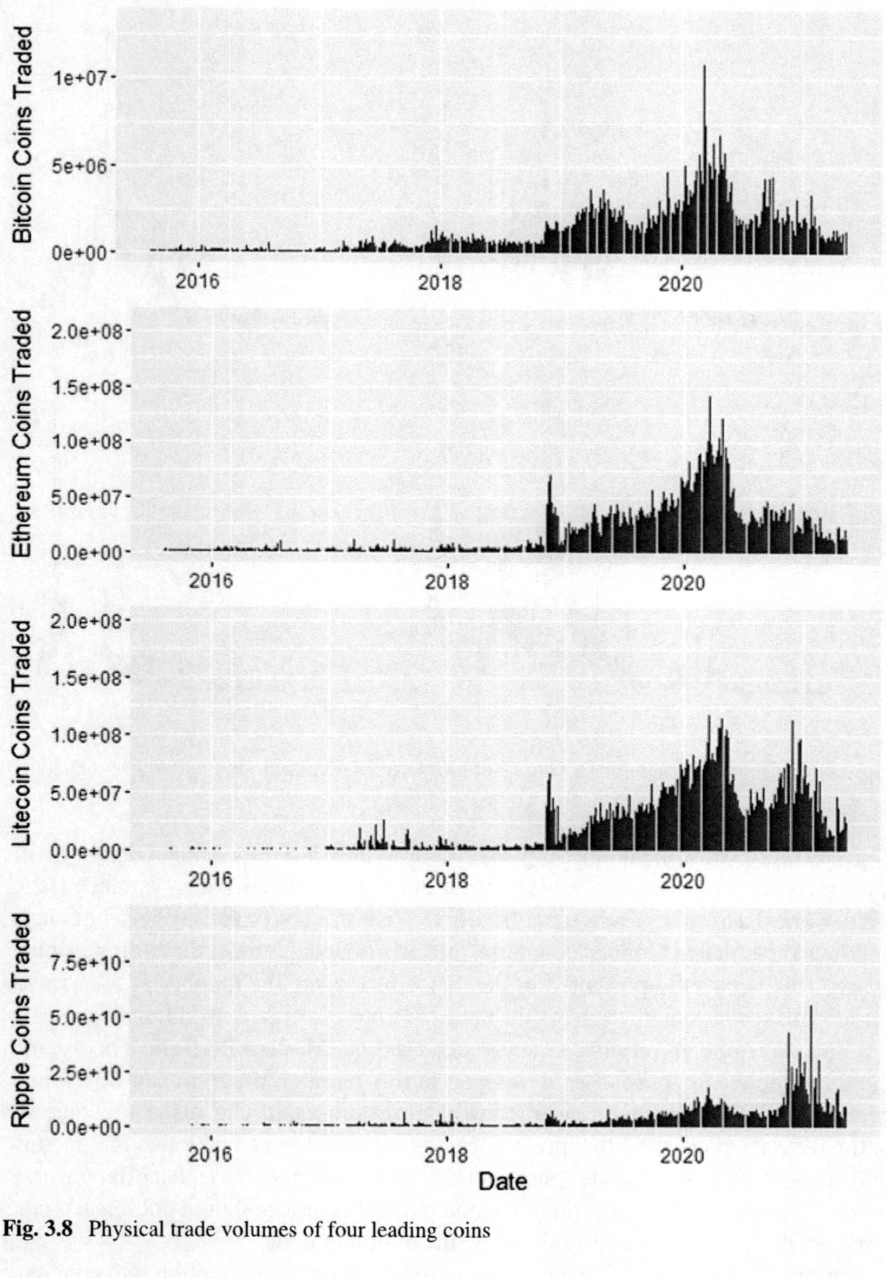

Fig. 3.8 Physical trade volumes of four leading coins

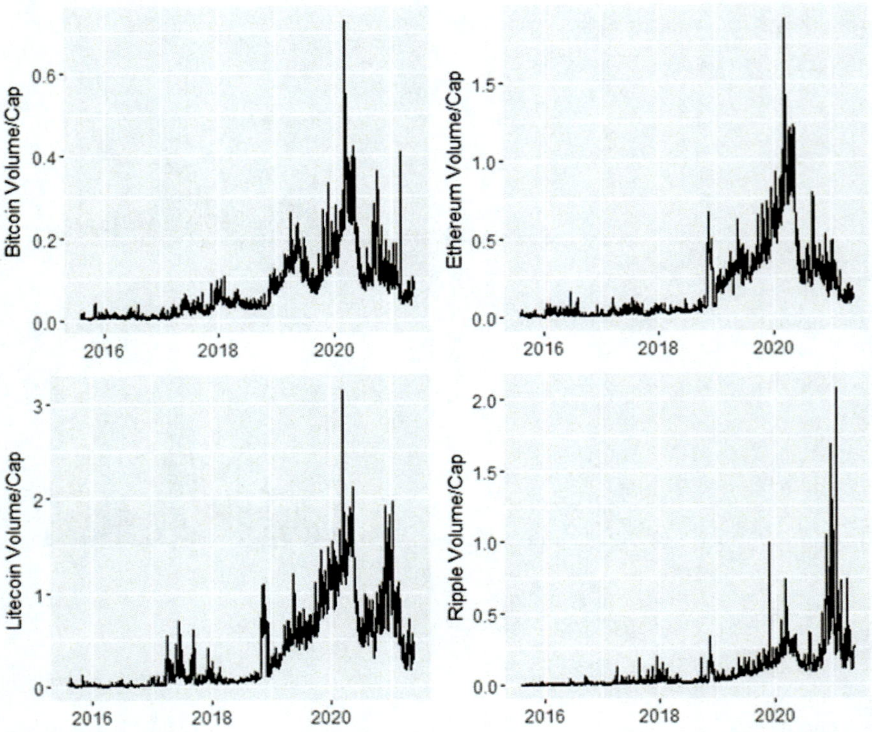

Fig. 3.9 Closing prices and trading volume of four selected cryptocurrencies, Q4.2015–Q2.2021

Essentially, this mean moving away from the two-dimensional analysis of the mean–variance framework, and into a three-dimensional mean–variance-trade framework. Return is now a function of two risk drivers (volatility and liquidity) and thus the efficient frontier now turns into an efficient plane in three-dimensional space. This is visually presented in Fig. 3.10. Increasing the number of assets will serve to make the construction of this plane even more accurate. Additionally, the 3D risk space is quite an intuitive extension of classical Modern Portfolio Theory and the same tools and approaches developed in this research program can be applied here, as well. Within this new framework investors again can make a choice of optimal assets given their risk preferences—only now have to take two risk factors into account. There is also the concept of dominated assets—those that offer a worse return for given volatility and liquidity, and rational investors should not select them. Graphically, dominated assets fall below the optimum plane.

For practical purposes, it is often useful to turn this visual intuition into concrete number for analytical purposes. Using Eq. (3.3), the adjusted Sharpe ratio can be easily calculated and rescaled accordingly. The worst performers according to S_A are the Vechain and Tron coins, and the best tradeoffs between return, risk, and liquidity is found in the Bitcoin Cash and Polkadot coins over the reference period. Naturally, the precise numbers will fluctuate as the sample is increased and the analyst will have

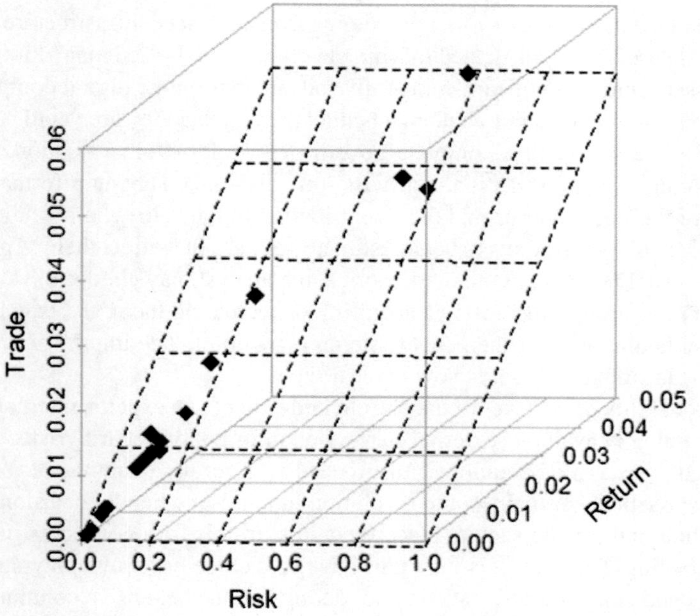

Fig. 3.10 Selected cryptocurrencies and the efficient plane in a 3-dimensional risk space

to frequently recalculate risk metrics. If needed, other traditional risk metrics may be corrected for trade volume, thus obtaining adjusted betas, Value at Risk, Expected Tail Loss, and others. The key insight from this exercise is the need to find the root cause of pertinent risks in innovative digital assets and rigorously include it in the risk analysis.

3.6 Conclusion

The emergence of novel digital assets is thrilling news for investors, with some pundits excitedly exclaiming that cryptocurrency is eating the world. In fact, the proliferation of many new information and digital assets provides vast opportunities for investment but also brings to the fore previously unknown risks. Digital transformation has given the world information assets and hugely inflated their importance. A key issue is that they are not yet fully understood, and thus individuals, companies and states have to be cognizant of how to calculate their values and estimate the likelihood of compromise. Those are the uncharted territories of managing risk in digital assets.

This chapter has shown a broad overview of those new types of assets, dividing them into two broad classes—information assets and digital financial assets. The former group includes software, hardware, physical, and service assets and forms the backbone of the new digital economy. Ranging from advanced infrastructure through extensive datasets to sophisticated machine learning models (and many more) information assets confer significant competitive advantages on the digital companies in the new economy. The major challenge behind them is that they are yet to be treated as full-blown assets with a complete governance cycle—from valuation, through accrual, audit, use, amortization, and write-off (disposal). Having a formal understanding of their actual price and their contribution to the value chain would enable individuals and organizations to better estimate the risks that affect them. Apart from taking the usual recourse to the market price, the analyst may choose to value them independently through the costs of acquiring or recreating them, the damage done if the asset is unusable, or the benefits stemming from it. The impact of a risk that affects this information asset is now in clear view.

The probability of occurrence may be elicited either from experts, internal data, or from vast publicly available external surveys on information security risks. In short, the risk analysis is fully equipped to proceed to rigorous quantitative modeling. Only a last challenge remains—the revolution in mindset whereby decision-makers in the public and private sectors place their trust in this formal process instead of their gut feeling. Risk analysis may thus have to include not only applying elegant valuations and sophisticated statistical modeling but also extensive communication that effectively ensures stakeholder buy-in.

In parallel to the proliferation of information assets, financial assets have gone digital, too. Traditional, and well-loved valuables such as money have moved to the digital world and new classes of assets have appeared spurred by new technologies such as the cryptocurrencies (coins), tokens, and cryptographic securities. Novel business models have spurred tokenized tangibles or intangibles such as the Non-Fungible Tokens (NFTs) that have rised in prominence. In a few years any discussion on the specifics on these assets as they stand today may be hopelessly outdated. This is why the risk analyst is well advised to look for basic drivers of risk in any novel asset, uncover the most important ones and create an appropriate ranking. The emergence of innovative digital currencies is a case in point—market liquidity looms as a large concern in the time immediately following their release and traditional metrics have to be adjusted accordingly.

It is not amiss to expect that rapid technological development will bring new opportunities to make decisions, produce, act, and invest in the digital world. More advanced forms of Artificial Intelligence will probably have a role in driving this process. The likely result is an increase in the risk exposure of individuals and organizations and a trend towards facing ever more sophisticated technogenic risks. Risk analysis becomes even more important in such an environment and the risk analyst is called upon not merely to apply traditional best practices but to obtain both a deep understanding and an appreciation of the underlying digital realities in order to perform a rigorous and useful risk evaluation.

References

1. Agrawal, V.: A comparative study on information security risk analysis methods. J. Comput. **12**(1), 57–67 (2017)
2. Ali, O., Shrestha, A., Chatfield, A., Murray, P.: Assessing information security risks in the cloud: a case study of Australian local government authorities. Gov. Inf. q. **37**(1), 101419 (2020)
3. Allodi, L., Massacci, F.: Security events and vulnerability data for cybersecurity risk estimation. Risk Anal. **37**(8), 1606–1627 (2017)
4. Berentsen, A., Schär, F.: The case for central bank electronic money and the non-case for central bank cryptocurrencies. Fed. Reserv. Bank St. Louis Rev. (Second Quarter), 97–106 (2018)
5. Bordo, M.D., Levin, A.T.: Central bank digital currency and the future of monetary policy (No. w23711). National Bureau of Economic Research (2017)
6. Borek, A., Parlikad, A.K., Webb, J., Woodall, P.: Total information risk management: maximizing the value of data and information assets. Newnes (2013)
7. Catalini, C., Gans, J.S.: Initial coin offerings and the value of crypto tokens (No. w24418). National Bureau of Economic Research (2018)
8. Chen, P.S., Yen, D.C., Lin, S.C.: The classification of information assets and risk assessment: an exploratory study using the case of c-bank. J. Glob. Inf. Manag. (JGIM) **23**(4), 26–54 (2015)
9. Conley, J.P.: Blockchain and the economics of crypto-tokens and initial coin offerings. Vanderbilt University Department of Economics Working Papers (17-00008) (2017)
10. Cram, W.A., Proudfoot, J.G., D'arcy, J.: Organizational information security policies: a review and research framework. Eur. J. Inf. Syst. **26**(6), 605–641 (2017)
11. Fernandes, K., Vinagre, P., Cortez, P.: A proactive intelligent decision support system for predicting the popularity of online news. In: Portuguese Conference on Artificial Intelligence, pp. 535–546. Springer, Cham (2015)
12. Fitzgerald, S., Jimenez, D.Z., Findling, S., Yukiharu Yorifuji, Y., Kumar, M., Lianfeng Wu. L., Giulia Carosella, G., Sandra Ng, S., Robert Parker, R., Carter, P., Whalen, M.: IDC FutureScape: Worldwide Digital Transformation 2021 Predictions. IDC (2020)
13. Giat, Y., Dreyfuss, M.: A composite risk model for optimizing information system security. In: Optimizing Data and New Methods for Efficient Knowledge Discovery and Information Resources Management: Emerging Research and Opportunities, pp. 74–97. IGI Global (2020)
14. Gronwald, M.: How explosive are cryptocurrency prices? Financ. Res. Lett. **38**, 101603 (2021)
15. Henderson, M.T., Raskin, M.: A regulatory classification of digital assets: toward an operational Howey test for cryptocurrencies, ICOs, and other digital assets. Colum. Bus. L. Rev. 443 (2019)
16. Herrmann, D. S. (2007). *Complete guide to security and privacy metrics: measuring regulatory compliance, operational resilience, and ROI*. Auerbach Publications.
17. Hu, B., Zhang, Z., Liu, J., Liu, Y., Yin, J., Lu, R., Lin, X.: A comprehensive survey on smart contract construction and execution: paradigms, tools, and systems. Patterns **2**(2), 1–51 (2021)
18. Hubbard, D.W.: The Failure of Risk Management: Why It's Broken and How to Fix It. Wiley (2020)
19. Huckle, S.J., White, M., Bhattacharya, R.: Towards a post-cash society: an application to convert fiat money into a cryptocurrency. First Monday (2017)
20. Ikegami, K., Kikuchi, H.: Modeling the risk of data breach incidents at the firm level. In: International Conference on Innovative Mobile and Internet Services in Ubiquitous Computing, pp. 135–148. Springer, Cham (2020)
21. Jaiswal, M.: Risk analysis in information technology. Int. J. Sci. Res. Eng. Dev. (IJSRED). ISSN 2581-7175 (2019)
22. Joshi, C., Singh, U.K.: Information security risks management framework–a step towards mitigating security risks in university network. J. Inf. Secur. Appl. **35**, 128–137 (2017)
23. Liu, C., Wang, H.: Crypto tokens and token offerings: an introduction. Cryptofinance and Mechanisms of Exchange, pp. 125–144 (2019)
24. Massad, T.G.: It's time to strengthen the regulation of crypto-assets. Econ. Stud. Brook. 34–36 (2019)

25. Milos, D., Gerasenko, V.: Prospects for the development of digital financial assets. Univ. Econ. Bull. **44**, 56–63 (2020)
26. Ortega, J.B.: Criptocurrencies, Fiat Money, blockchains and databases. arXiv preprint (2020). arXiv:2002.08466
27. Parsons, K., Calic, D., Pattinson, M., Butavicius, M., McCormac, A., Zwaans, T.: The human aspects of information security questionnaire (HAIS-Q): two further validation studies. Comput. Secur. **66**, 40–51 (2017)
28. Peltier, T. R. (2005). Information security risk analysis. CRC press.
29. Posey, C., Roberts, T.L., Lowry, P.B., Bennett, R.J., Courtney, J.F.: Insiders' protection of organizational information assets: development of a systematics-based taxonomy and theory of diversity for protection-motivated behaviors. Mis Q. 1189–1210 (2013)
30. Raskin, M., Yermack, D.: Digital currencies, decentralized ledgers and the future of central banking. Research Handbook on Central Banking. Edward Elgar Publishing (2018)
31. Roth, J., Schär, F., Schöpfer, A.: The Tokenization of assets: using blockchains for equity crowdfunding. SSRN 3443382 (2019)
32. Schär, F.: Decentralized finance: on blockchain-and smart contract-based financial markets. SSRN 3571335 (2020)
33. Shamala, P., Ahmad, R.: A proposed taxonomy of assets for information security risk assessment (ISRA). In: 2014 4th World Congress on Information and Communication Technologies (WICT 2014), pp. 29–33. IEEE (2014)
34. Shatohina, J.I., Kochetkov, A.V.: Actual issues of risk management in the market of digital financial assets. In: International Conference on Economics, Management and Technologies 2020 (ICEMT 2020), pp. 323–327. Atlantis Press (2020)
35. Stoneburner, G., Goguen, A., Feringa, A.: Risk management guide for information technology systems. NIST Spec. Publ. **800**(30), 800–830 (2002)
36. Toygar, A., Rohm, C.E., Jr., Zhu, J.: A new asset type: digital assets. J. Int. Technol. Inf. Manag. **22**(4), 7 (2013)
37. Trautman, L.J.: Virtual art and non-fungible tokens. SSRN 3814087 (2021)
38. Tsaregorodtsev, A.V., Kravets, O.J., Choporov, O.N., Zelenina, A.N.: Information security risk estimation for cloud infrastructure. Int. J. Inf. Technol. Secur. **10**(4) (2018)
39. Vlasov, A.V.: The evolution of E-money. Eur. Res. Stud. **XX**(1), 215–224 (2017)
40. Wilson, R.M., Stenson, J.A.: Valuation of information assets on the balance sheet: the recognition and approaches to the valuation of intangible assets. Bus. Inf. Rev. **25**(3), 167–182 (2008)

Chapter 4
Networks of Risk

4.1 Introduction

In 1996 two Stanford Ph.D. students would stumble upon an insight that was deemed revolutionary at the time, considered trivial today, and eventually made them billionaires. Larry Page and Sergey Brin were working on a research project for a search engine within the university, when they realized that web content may be subject to a hierarchy of popularity, and more popular websites are likely to be more relevant and useful. Page and Brin thus decided to measure popularity as the number of links connecting to a given site and rank it accordingly. The PageRank algorithms was thus born which later served as the backbone of the Google Search Engine, and the main profit driver of Google Inc., later renamed Alphabet. Stanford University received rights to the patent and licensed it back to the inventors for a meagre 1.8 million shares in Google, later sold for $336 Million. Page and Brin's revolution, in short, was to discern that like many other real-life phenomena, the internet was a network with tractable typology that yields itself to analysis.

Network formation is basic to human interactions. A complex socio-economic system may often be represented as a network with actors (human, organizations, etc.) serving as the network nodes, and all their interactions (trade, information diffusion, contagion, etc.) being the connecting edges. The network structure is ubiquitous in many domains ranging from food network, electric power grids, cellular and metabolic networks, the internet, telephone networks, scientific citations, executive board membership, and many others.[1] Our economies and cities are no less structured as a network than our personal relations and critical infrastructures.[2] How different nodes are connected in such systems has crucial implications for their functioning, stability, resilience, and most importantly—the risks generated within.

Traditional risk metrics are preoccupied with finding the fundamental underlying risk of a single actor or asset, and then aggregating those to gain an understanding

[1] Strogatz [45].

[2] West [52].

A. Gerunov, *Risk Analysis for the Digital Age*, Studies in Systems, Decision and Control 219, https://doi.org/10.1007/978-3-031-18100-9_4

of the overall risk profile. This approach misses the fact that a large part of the risk may not be due to the fundamentals of the actor or asset but rather by its place in the network, and its ability to generate and absorb shocks that propagate throughout the network. Interconnectedness in the modern economy spells an increasing dependence on a large network of counterparties (customers, vendors, partners and so on), whose well-being is directly relatable to the well-being of the individual agent or position. Issues such as network contagion come front and center when discussing both systemic risk, and the new realities of individual risk.

An example from epidemiology further clarifies the point. Human societies are network formations whereby individuals (nodes) communicate with other individuals. In the case of unexpected global pandemics such as the one that erupted in 2020, the key questions revolve around risk of spread and contagion. Traditional risk management approaches will tend to take each individual and estimate their probability of contracting the disease, marking it as the relevant risk level. However, the probability depends not only on this individual and their fundamentals (health status, immune response, anti-bodies, overall level of infection in the region, etc.) but also on the intensity of their interaction with other people and other people's fundamentals and susceptibility. The network typology thus informs the analyst of the structure of transmission, the likelihood of contraction, and the influence of each individual upon others in their vicinity. In this way one obtains a much clearer and nuanced picture of risk and is able to target risk management efforts to where they matter most—likely towards the super-spreaders.

The example of highly interconnected banking readily jumps to mind as an apt analogy. During the global financial crisis that started in 2007–2008, the US financial system featured a set of highly connected banks with large mutual exposures. As one of them defaulted, the balance sheets of its creditors deteriorated. Some of them worsened to such extent that they themselves risked insolvency, thus endangering the financial health of their own counterparties. Such contagions have the ability to produce a domino effect with grave repercussions for the whole system. As a result, the network structure of the financial sector is increasingly acknowledged as a major risk driver which starts to attract major research attention.[3]

This chapter starts by looking at different networks and their potential to dampen or (most often) amplify risk. It also reviews different types of network formation mechanisms and their resulting topologies in the hope that these will inform a more sophisticated view of real-life risk exposures. Despite network analysis still being a relatively young field, one cannot escape from the imperative to understand networks in the increasingly interconnected world we live in. Formal metrics of the network structure are introduced with clear recourse to what they mean for the risk manager. Finally, the focus shifts to the actual construction of networks with empirical data leveraging a Bayesian approach.

[3] Gai and Kapadia [18].

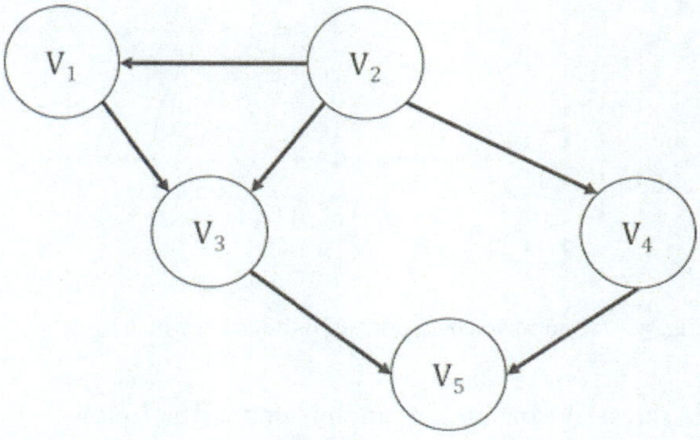

Fig. 4.1 A simple directed network with five vertices

4.2 Networks and Risk

Networks have always been ubiquitous but only recently rose in prominence and sparked interest in researchers and practitioners. This is not due to the fact that they did not exist or draw attention—the biological networks that weave the fabric of life have always been around, and attention to some human networks such as Europe's royal dynasties has been amply displayed. Still, the twentieth century and the early twenty-first century provide unique foundations for the study of networks. The proliferation of cheap computing, the ability for rapid sharing of large datasets, and most notably—the growing availability of network maps spurred the study of those connections that later came to be known as network science.[4] The increasing quantification and use of rigorous mathematical methods allow the analyst to formally investigate the network topology and quantify the implications of network dynamics for risk management. Graph theory is usually the mathematical tool of choice when investigating network structures, and the next section provides a brief overview of a few important highlights.

4.2.1 *System Structure and Risk*

The view of the network is intuitive—one easily pictures that as being composed of a number of nodes, with some of them being connected to one or more of the other. The greater significance behind this sketch is that every node is an agent, activity, or artefact of interest; and every connecting line is a form of interaction, exposure, or flow. This logic is graphically presented in Fig. 1.3, with nodes (or vertices) denoted as V_i (Fig. 4.1).

This structure can be more formally defined as a graph, which is usually denoted by G. The graph is an ordered triplet of vertices (or nodes), V, edges (or lines), E, and a function f that maps each line from the set E to two unordered elements (vertices)

[4] For an excellent introduction to networks, the reader is referred to Barabási [2] and Newman [36].

	V_1	V_2	V_3	V_4	V_5
V_1		1	1		
V_2	1		1	1	
V_3	1	1			1
V_4		1			1
V_5			1	1	

$$\rightarrow \begin{pmatrix} 0 & 1 & 1 & 0 & 0 \\ 1 & 0 & 1 & 1 & 0 \\ 1 & 1 & 0 & 0 & 1 \\ 0 & 1 & 0 & 0 & 1 \\ 0 & 0 & 1 & 1 & 0 \end{pmatrix} \rightarrow$$

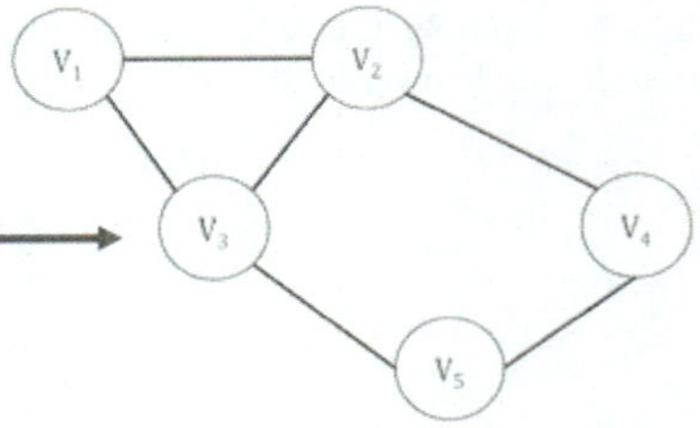

Fig. 4.2 Undirected network and its adjacency matrix

of the set V. In short, a graph is defined as follows:

$$G := (V, E, f) \tag{4.1}$$

If the mapping function is obvious or unimportant to the problem, oftentimes the graph may be defined by just the vertices and edges $G := (V, E)$. Usually the correspondence between real-life problems and the graph representation is straightforward as nodes tend to represent real-life entities of interest, while edges—their interactions. In a social media network such as Facebook the nodes may be individual users while the vertices may be the interactions they have, such as likes, comments, shares, and many others. Those edges may be directed or undirected. The former corresponds to situations where there is a flow (or action) from one entity to other but not in the oppositive direction. The latter correspond to reciprocity. For example, in a network of friends a line may denote a flow of information. If there is a directed edge from V_1 to V_2 that means that V_1 shares information with V_2 but V_2 does not share with V_1. An undirected edge can denote dialogue and mutual sharing, instead.

A network can be formally described by means of an adjacency matrix—a matrix that has vertices in both rows and columns and keeps track of the existence of lines between the vertices. Assuming a network of 5 vertices such as the one in Fig. 1.3, one can represent it in a 5-by-5 matrix whereby the line between V_i and V_j is represented by the number 1 in ijth position in the matrix. The undirected version of the network in Fig. 1.3 and its corresponding adjacency matrix is presented visually in Fig. 4.2.

In a similar vein one can represent directed lines in the adjacency matrix. Should a directed link exist from V_j to V_i but not vice versa, there would be a 1 in the jith position, but 0 in the ijth position of the matrix. This information can readily be contained not only in an adjacency matrix but also in similar structures such as the edge list (which enumerates the nodes that are connected by an edge) or an adjacency list (essentially a list of nodes, followed by the nodes they are connected with). Either way, the network structure is defined by vertices and edges, and the analyst needs structured information for both in order to perform risk estimation tasks.

Network typology may be a crucial tool for understanding risk, and how it is propagated through a complex socio-economic system. Looking at the world's largest Ponzi scheme may further illustrate this point. Enter Bernard Madoff.[5]

[5] Kirchner [26].

Beginning in 1960 with trade in penny stocks, Bernard (Bernie) Madoff grew his investment company into a Wall Street behemoth—a market maker and a respected financial company under the name of Bernard L. Madoff Investment Securities. Madoff himself was seen an exceptionally bright investor, a philanthropist, a community pillar, and even served as the non-executive Chairman of the NASDAQ stock exchange.[6] There was only a slight problem—Madoff was in fact operating a Ponzi scheme through an obscure network of financial transactions with a large number of suspecting and unsuspecting counterparties under questionable supervision from partners, auditors and regulators.[7] The eventual losses to investors reached USD 64.8 Billion, while Mr. Madoff and his family enriched themselves handsomely. The story ended as most Ponzi schemes do, and as of the time of writing Bernard Madoff is serving a 150 year prison sentence.

The key question, however, remains, and it is how such a Ponzi scheme would be possible to sustain over a period of three decades. A glimpse into this may be given by understanding Mr. Madoff's place in the network and mapping out the money flows from his investment company to other (un)witting counterparties. This is mapped using data from The Network Thinkers, as provided by Lin Freeman,[8] and visualized in Fig. 4.3. The central node of the network is Bernard Madoff Investment Securities, and every link is a monetary transaction. A few key insights readily spring from the network visualization.

First, Madoff was dealing with a large number of loosely connected counterparties so that no single agent in the system but him would have a complete overview of the financial transactions. Second, many of the counterparties are not among the largest financial institutions on the market which means that they may not have sufficient resources, expertise and incentives to perform extensive due diligence. Third, the network consists of one strongly connected central component which is responsible for a large number of transactions, thus enabling trust-building and recurring profitability for the counterparties. Fourth, the reach of the network is surprising in scope, eventually reaching banks and funds from Europe to Asia. In short, the network presents a very sophisticated and opaque financial operation that does not easily lend itself to tracing and analysis even by the most diligent of compliance officers or auditors.

One can further analyze the network by looking at its stylized structure presented in Fig. 4.4, with Madoff's operation again at the center of the network. We clearly see that Madoff's investment company is central to this network and it initiates the flow of money to the boundary notes. The father away the boundary node is (i.e. the longer the path to it), the more difficult it is to connect transactions to their source. Furthermore, international entities are subject to different kinds of regulation and audit requirements, and while the network of money flow is monolithic for Madoff, it is a puzzle of different jurisdictions for the regulators. The key insight here is that

6 Krugman [28], Hurt [24].

7 Gregoriou and Lhabitant [20].

8 Freeman [16].

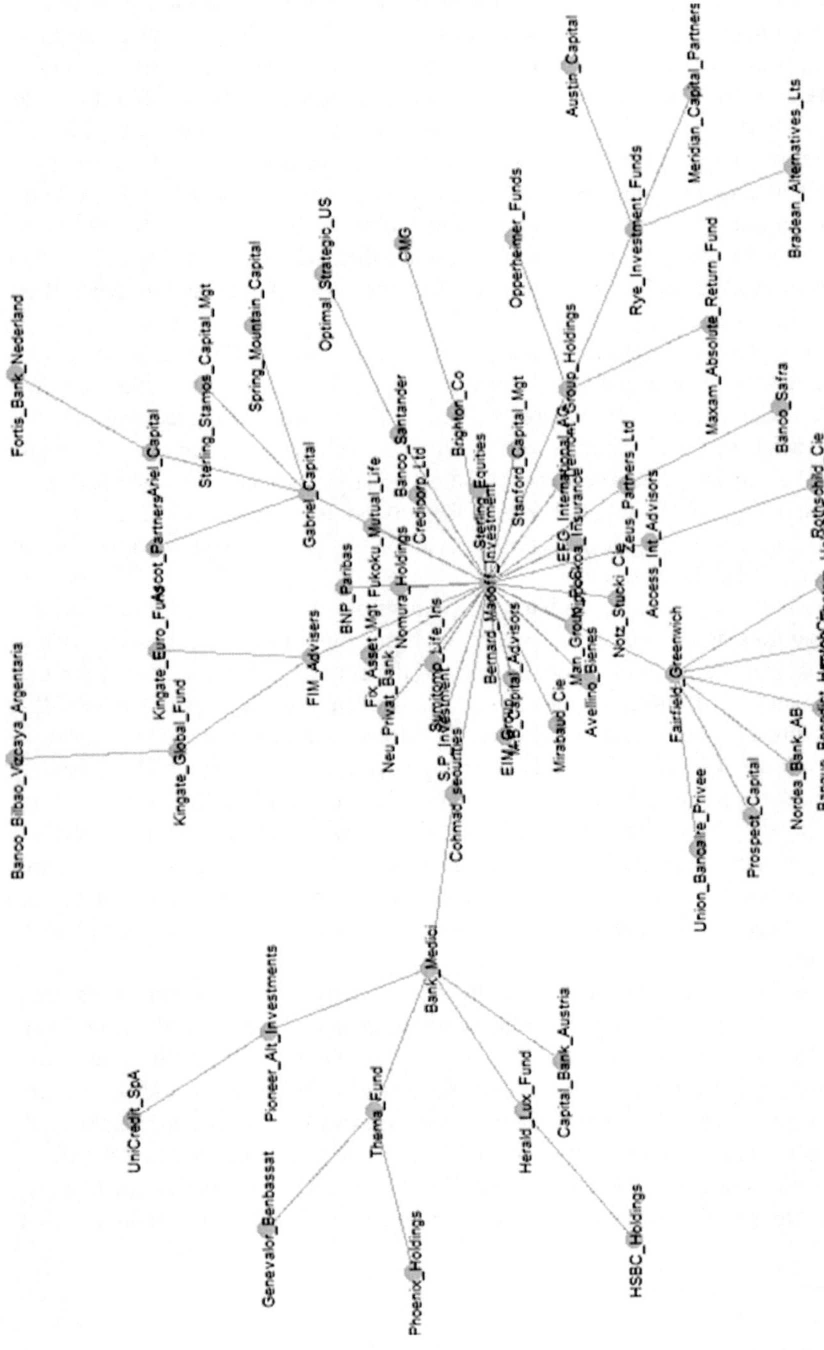

Fig. 4.3 Network of Bernard Madoff's financial dealings

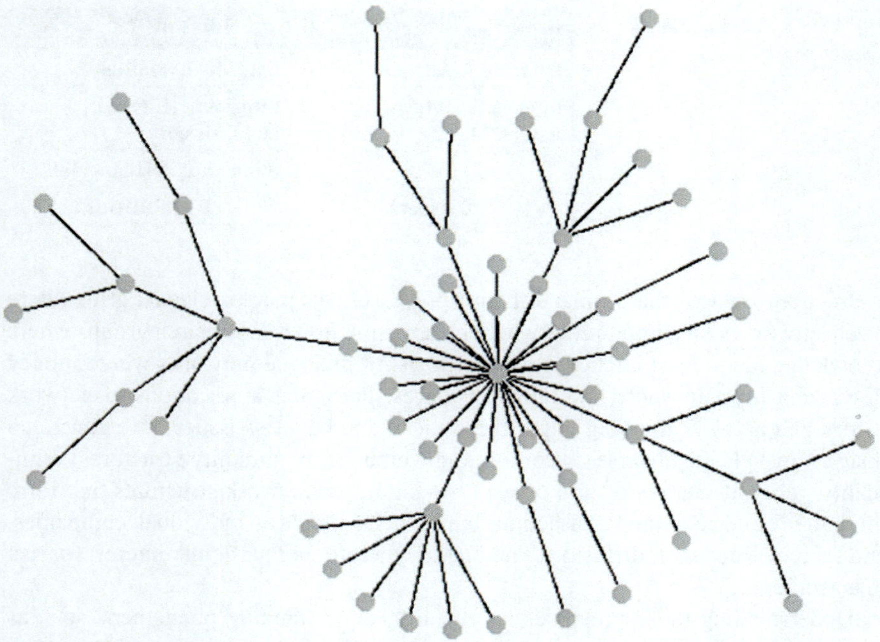

Fig. 4.4 Stylized network of Bernard Madoff's financial dealings

the position of a node within the network is of crucial importance, and that it, together with the flows between nodes can be meaningfully used to map and analyze risk.

4.2.2 Network Diffusion

The key benefit of viewing the world as a network is the ability to conceptualize not merely the agents (nodes) but, more importantly, the interactions between them (the vertices). The structure of flows essentially represents diffusion in the network and underpins the importance of interconnectedness—in fact, if the flow was zero, then no network would be needed to understand the system. One of the earliest applications of network diffusion was to understand pandemics. In an interconnected web of individuals, a viral disease "flows" from one individual to those closest to him or her, with a given probability.[9] The newly infected in turn can transmit the disease to those closest to them, and so on, until the viral diseases transmits across the network. The average number outgoing viral transmissions is the average number of diseased people per individual. That is precisely the basic reproduction number of the disease, or in network parlance—the average out-degree of the network.

[9] Stattner and Vidot [44].

Fig. 4.5 A Simplified Bank balance sheet

Assets	Liabilities
External Assets, EA_i	External Liabilities, EL_i
Interbank (Internal) Assets, IA_i	Interbank (Internal) Liabilities, IL_i
	Bank Capital (Equity), K_i
Σ Assets	**Σ Liabilities**

It is quite obvious that similar diffusion processes take place in electric grids where electricity flows, in communities where information flows, in research groups where knowledge (or at least citations) readily flows, in financial networks where money flows, and in many more additional instances. Particular applications of network theory and network thinking to pertinent social and business issues have generated insights in fields as diverse as corporate governance, community structures, criminality, religious networks, and others.[10] In short, the network structures that form modern life and economic transactions have crucial effects on individual, companies, and societies, and their diffusion dynamics are therefore of particular interest for risk management.

Understanding those processes is crucial to understanding phenomena such as financial contagion and the propagation of financial risk through a complex system. This can be illustrated by taking recourse to a simple contagion model, proposed by Gai and Kapadia.[11] They posit a network of n connected financial intermediaries (e.g. "banks"), represented by nodes, whereby the unsecured exposure of bank i to other banks defines the vertices of the network. The links are directed as a bank can have both assets (borrowing to others) or liabilities (lending to others). Each bank i has a number of lending links j_i and a number of borrowing links k_i. Since each lending link for one bank is a borrowing link for another, then the average number of lending and borrowing links must be equal. This number is an indicator of the system's connectivity or, alternatively, its average degree and its denoted by z in the model.

Each individual bank's balance sheet is composed of two types of assets and liabilities—internal and external to the network. External assets (EA_i) include ones such as mortgages and corporate loans, while external liabilities (EL_i) are primarily composed of customer deposits. The model does not look at external positions but focuses on how the dynamics of connections internal to the network may bring about financial contagion. A stylized balance sheet of a given bank (node) with respective notations is graphically presented in Fig. 4.5.

Since every bank's internal liability (IL_i) is another bank's internal asset (IA_i), the sum of all unsecured interbank assets is equal to the sum of all liabilities. While this is true at the systemic level, it does not hold at the individual bank level as any single bank may run a surplus or a deficit. The latter ones may be particularly susceptible

[10] Scott [39].

[11] Gai and Kapadia [18].

to risk. Therefore, if a given fraction of the bank's counterparties default (denoted as ϕ), then this bank loses what the defaulted counterparties own and has to cover the loss with what remains on the balance sheet. Thus, the solvency condition is that the sum of remaining internal assets and its external assets is larger that the sum total of all outstanding liabilities. The conditions is thus as follows:

$$(1-\phi)IA_i + EA_i - IL_i - EL_i > 0 \tag{4.2}$$

One can express the shareholder equity K_i as follows:

$$K_i = IA_i + EA_i - IL_i - EL_i \tag{4.3}$$

Substituting Eq. (4.3) into Eq. (4.2) we reach the following solvency condition:

$$\frac{K_i}{IA_i} > \phi \tag{4.4}$$

Equation (4.4) has an intuitive economic interpretation—as one of the bank's counterparties defaults, the loss needs to be covered by the shareholder equity. As more and more banks default, the losses become larger than the amount of equity, and thus this struggling bank also defaults. This increases in turn the losses to its own lenders and calls for ever more of their own shareholder capital to cover those. The pressure increases and if one of the counterparties defaults then the risk for its own partners skyrockets and tests their own ability to absorb loss. Sufficiently weak banks fall like pieces of domino and their downfall worsens the balances of stronger ones that may not be able to withstand such large losses. This way, contagion spreads throughout the system.

This process is analytically tractable within the model, recalling that each bank i has j_i lending links. One can assume for simplicity that interbank assets are evenly distributed across banks. In this case $\phi = 1/j_i$, which when substituted into Eq. (4.4) yields the condition for contagion to spread. For this to happen there must be at least one neighboring party to a defaulted bank for which the following holds:

$$\frac{1}{j_i} > \frac{K_i}{IA_i} \tag{4.5}$$

Imposing more assumptions on the model would make it a bit less general but would enable the analyst to obtain tractable analytical results on the diffusion process.[12] Supposing that the network is not randomly generated but rather that each bank is equally connected to exactly z bank (so $j_i = k_i = z$) gives an approximation of the model. Furthermore, if one assumes that all banks have identical balance sheets, this leads to the equality that $IA_i = IL_i$. Substituting this into Eq. (4.4) one reaches the following condition

[12] May and Arinaminpathy [32].

$$z < \frac{K_i}{I A_i} \tag{4.6}$$

Thus, the expression given in Eq. (4.6) is the tipping point for network contagion. If the network is sufficiently connected with values of $z \geq 1$, then any individual bank default leads to the default of all its counterparties, and so on until defaults propagate throughout the whole network. Conversely, if conditions (4.5) and (4.6) do not hold, then no network contagion takes place. While this conclusion holds under very stringent assumptions, its general message remains and carries well over to realistic networks. Namely, it is that even small changes in the network topology and its underlying features (connectivity and exposure) may lead to very different outcomes. It this vein, the conclusion from network theory is similar to the messages of complexity theory. This is hardly surprising as both endeavor to model and investigate the same underlying phenomenon—complex systems in society, nature, and business.

Model simulation gives other intriguing results in a similar direction.[13] This system is subject to two phase transitions, or alternatively—has two modes of operation. If the connectivity is too low, then contagion fails to spread. On the other hand, if connectivity is too high, then there is ample opportunity for risk diversification, as many connections mean that individual defaults are a tiny fraction of internal assets. In this case, contagion also fails to spread. In short, the network needs to be sufficiently connected to spread risk but insufficiently connected to diversify it. In this case alone is widespread diffusion a high probability. Simulations also show the so-called robust-yet-fragile property—this means that while the probability of contagion is very low as the network becomes more and more connected, the few cases of contagion eventually reach the whole network.

Moreover, while the highly connected network is impervious to most individual defaults, there are some nodes that may lead to complete collapse should problems occur there. This particular model can be extended to include not merely defaults but also other events that have repercussions on the quality of banks' balance sheets. Among those we should particularly note the inclusion of liquidity hoarding whereby banks withdraw their loans for precautionary motives,[14] and the inclusion of fire sales as financial agents seeks to liquidate assets at any price.[15] Such models can be extremely useful for risk management as they complement and deepen traditional stress testing of the financial sector and can serve as tools for policy formulation. Most notably, regulators and risk managers who leverage them are able to pinpoint particularly fragile parts of the system.

More generally, models of network diffusion give one both the idea of how risk flows throughout and system as well as an estimate of the riskiest nodes and the conditions of overall network failure. This gives risk analysis an additional level of depth compared to looking at traditional metrics at the level of individual agents and

---

[13] Watts [51], Gai and Kapadia [17].

[14] Gai et al. [19].

[15] Cifuentes et al. [11], May and Arinaminpathy [32].

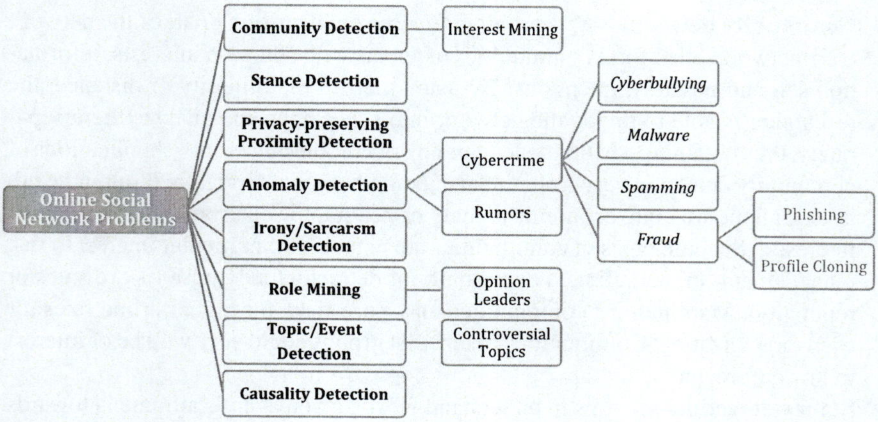

Fig. 4.6 Types of online social network analysis problems. *Source* Can and Alatas [8]

enables the study of correlated phenomena and agents in search to better understand cascading failures and their trigger events.

4.2.3 Social Networks

The two decades have seen the rapid emergence and exponential growth of another type of network structures that are mostly rooted in the digital environment. Those are the online social networks (OSNs) that nowadays connect billions of users and attract the attention of millions of companies. Online social networks are communities of people with common interests who enjoy common activities, have similar backgrounds, and form friendships among them[16]—either as a reflection of offline communication, or even without the need of prior real-life interaction. This structure can be aptly captured as a network, whereby individual agents (users) are the nodes, and their different interactions and connections (e.g. friendships, following, shares, likes, messaging, etc.) are the edges of the network. The topology of those social networks can be exploited for a wide set of research tasks and is already of interest in fields as diverse as entertainment, education, communication, commerce, and business (Fig. 4.6).[17]

Online social network analysis can be used to solve diverse problems that can be classified using a range of different criteria. For brevity we review just one of those typologies that include 21 large groups of tasks under eight large umbrella headings, following the typology of Can and Alatas.[18] The eight key problem domains are as follows:

[16] Schneider et al. [38].

[17] Kaur and Singh [25].

[18] Can and Alatas [8].

- **Community detection**—this problem focuses on identifying part of the network (sub-networks) that form communities of agents with common interests, information set, and possibly, intentions. By using metrics of similarity or distance, the researcher is able to divide the network into sub-graphs and analyze them separately, thus obtaining a better understanding of complex networks. Applications of community detection very from understanding human interaction, through health issues, all the way into economic activity or geographic distribution of features of interest.[19] Specific types of communities are naturally of particular interest to risk management. In marketing, a community of detractors may pose risk to sales or reputation. Marginalized communities may pose risks for higher crime rates, or even radicalization. Communities of terrorist groups could very well be of interest to law enforcement.
- **Stance detection**—it aims to understand what a given agent's attitude is towards a specific subject of interest (i.e. is the agent positive, negative, or neutral). It this vein this is rather similar to opinion mining and sentiment analysis, and is thus a useful task for social network aficionados and savvy marketers, but is generally of less interest to risk managers.
- **Irony and sarcasm detection**—in a similar vein this set of problems focuses on distinguishing between texts that are ironic in nature, and those that are not. While this is an interesting problem from the perspective of applied linguistics, it is currently rarely used for risk analysis. With the increasing growth of text analytics for risk management purposes, approaches to sarcasm detection may enhance the understanding of text and thus improve one's understanding of its implications.
- **Privacy-preserving proximity detection**—the proximity of nodes on the network is an important feature of its typology, and it can intuitively be transferred to actual real-life environments. Proximity services whereby users are able to allocate nearby users that are similar to them, or places or activities nearby (e.g. a restaurant or fitness center) lead to transmission of personal data. This data can then be obtained by malicious actors and used to criminal ends. Privacy-preserving proximity services aims to combine knowledge of the network topology with secure transmission channels to avoid that. This set of network problems can be a source of risk and a set of risk controls rather than methods for risk analysis.
- **Role mining**—this specific application is initially influenced by role theory in sociology and attempts to learn the role of each node in the network, based on both this node's characteristics, as well as the overall network topology. The most intuitive example is finding the influencers in the social network that will yield disproportionate ability to sway other agents' actions and thoughts. In a similar vein role mining can be used to find anomalous nodes that do not behave like the rest and take the cue that these are probably sources of risk. Understanding opinion leaders in their respective sub-network is a central issue when studying the diffusion of information and misinformation, and is particularly valuable in fields such as healthcare, marketing, and political communication.

[19] Tang and Liu [46].

- **Topic or event detection**—this line of inquiry started in communications research where it aimed to identify events and their representation in classic media. In the world of online social networks topic and event detection tasks usually leverage a large amount of data to automatically detect different events (e.g. an earthquake or a strike) and even be able to predict them. Early prediction of potentially catastrophic risks based on the collective wisdom of the online crowds is of obvious interest to risk analysis and risk management, and one surmises, automated event detection will only grow in importance.
- **Causality detection**—this research problem focuses on finding causal links between a set of explanatory variables and a certain outcome. OSNs are particularly suited to this type of problem as they allow the collection of data both on human behaviors and interactions, as well as on system designs. Research along those lines usually tries to connect behaviors in the network with actual results such as the link between social media opinions and activities on the one hand and voting patterns on the other. Much research has been done on the link between social media sentiment and economic indicators such as stock market prices. Causality detection studies are thus indispensable for both building risk management models as well as gaining a deeper understanding of the risk landscape.
- **Anomaly detection**—the search for anomalies in network data has turned into an important and fruitful task in the OSN research agenda.[20] It usually revolves around finding a dramatic change in the interaction (link) between two nodes, or a significantly different pattern of interaction as compared to the expected standard one. Those anomalies can be thought of exceptions, outliers, abnormalities, or contaminators. At any rate, the anomaly is a deviation from one's expectation and thus a potential risk. Anomalies may be due to a large variety of underlying processes—ranging from the more benevolent such as rumor detection, to more sinister ones such as cybercrime. The latter group contains activities such as injecting malware, spamming, cyberbullying, and fraud. The set of anomaly detection problems in social networks give solutions that can most readily be used by risk managers. In fact, fraud detection is one of the key tools to manage operational risks in digital transactions.

At any rate, online social networks have served to increase the exposure to risk in the digital economy, and to create links between online intention and real-world effect, ranging from stock market frenzies to religious radicalization. Thus, modeling and understanding social network dynamics may be a fruitful venue to gain further insight into overall individual, organizational, and systemic risk.

---

[20] Bindu and Thilagan [5].

4.3 Network Topologies

Once the risk manager starts seeing the world as a network, the key issue is to understand its underlying structure and see what implications for the risk landscape stem from that. Network science has been hard at work to propose ideas for the underlying drivers of network formation and the resultant network properties.[21] First passes at understanding networks included the rather stringent assumption that they are driven by random processes, whereby each node is connected to others by a random process with a certain probability. While this seemed like an intuitive first benchmark, it turned out that many real-life networks exhibit more complex properties than the simple random model would assume. In short, there must be a deeper underlying order to networks than simply chance. Thus, issues like network growth over time and preferential attachment came to the fore. More sophisticated network topologies empower the risk analyst to understand more complex underlying phenomena and predict key network dynamics such as diffusion of information, innovation, and contagion. To this end, the section briefly reviews the theory behind different types of network topologies and explores the results of possible network failures.

4.3.1 Random Erdös-Renyi Networks

The simplest network topology is that of the random network. Such a network may be formed when a group of agents (e.g. people) that have no connection are placed in a situation to interact. Strangers talk to strangers with no prior preference and a network of conversation emerges throughout the process. Such a random network is a simple object with some number of vertices (or nodes), N, and a probability p that a link (or edge) E is going to form between them. One of the earliest definitions is that of Gilbert, and it defines this graph as follows:

$$G = G(N, p) \tag{4.7}$$

Alternatively, this can also be defined as a graph with N nodes, and E randomly placed edges: $G(N, E)$. As Gilbert was introducing the idea of random networks, a couple of scientists—Pál Erdős and Alfréd Rényi—published a sequence of eight extremely influential papers on random networks over the period 1959–1968. Their impact was so strong that random networks came to called after them and to this day are known also as Erdös-Renyi Networks. Constructing such a network is a straightforward task, once connection probability p is defined. The algorithm is as follows:

1. Start with a pair of the N unconnected nodes;

---

[21] Barabási [2].

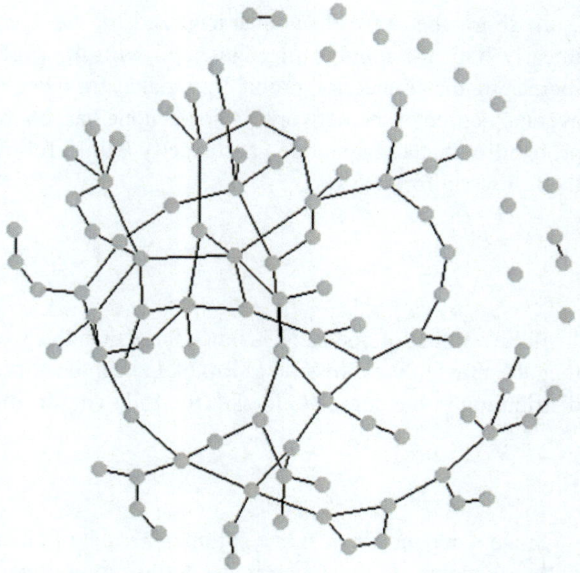

Fig. 4.7 Random (Erdös-Renyi) network with 100 vertices and 100 edges

2. Generate a random number between 0 and 1, and check if this number is large or smaller than p;
3. If the number is larger than p—connect the nodes, and if it is smaller—do nothing;
4. Repeat this procedure for every dyad of nodes, or in total $\frac{N(N-1)}{2}$ times.

The resulting network (an example is shown in Fig. 4.7) is a random one and it can be aptly characterized by the probability that it has exactly a number of L edges. This probability follows a Binomial distribution and can be characterized as follows:

$$p_L = \binom{\frac{N(N-1)}{2}}{L} p^L (1-p)^{\frac{N(N-1)}{2}-L} \tag{4.8}$$

The expected number of edges is thus:

$$E[L] = \sum_{L=0}^{\frac{N(N-1)}{2}} L p_L = p\frac{N(N-1)}{2} \tag{4.9}$$

From the expected number of edges one can easily calculate the expected average degree of the network $\overline{k}$, which is given by the following:

$$\overline{k} = \frac{2E[L]}{N} = p(N-1) \tag{4.10}$$

In short, the network as characterized by the average number of links grows linearly with the number of edges, and with the probability of them connecting. Increasing the connection probability makes for a denser network by increasing the average degree of the network, i.e. each node has, on average, more links. Thus, the probability that a given node i has exactly k links follows a binomial distribution of the following form:

$$p_k = \binom{N-1}{k} p^k (1-p)^{N-1-k} \tag{4.11}$$

If the number of nodes, N, is much large than the average number of connections, then the unwieldy expression in Eq. (4.11) can be approximated by using a Poisson distribution in the degree k. It takes the following form:

$$p_k = e^{-\bar{k}\frac{\bar{k}^k}{k!}} \tag{4.12}$$

Since it depends on just one parameter and it often calculated in software utilities without recourse to the equation itself, the expression in (4.12) is often the preferred way to characterize the properties of a random network. Essentially this is both the greatest strength and the worst weakness of the Erdös-Renyi network—it is truly random, and while this provides a valuable benchmark for real-life networks, its topology rarely resembles them. The main issue lies in the fact that this is an essentially static representation, while actual networks change over time. Moreover, networks in reality are driven by the goals of the agents therein, the decision rules and the decision architectures, which have crucial impact on the links formed. As an off-the-cuff example, a random network will say there is an equal probability of you knowing your own family as it is of knowing any other family in the city. To tackle those issues, a host of more sophisticated network models are offered which are often inspired by empirical regularities and deepened understanding of social and economic processes.

4.3.2 Growth and Preferential Attachment in Networks

The key empirical observation that defies the random network model is the presence of a so-called giant component in networks. This phenomenon refers to the emergence of a highly connected subgraph in the network whereby some nodes have a noticeably higher degree than those at the edges of the network. In short, one observes that a few agents have an extremely high number of interactions, while most do not. This immediately calls to mind Pareto's principle as well as numerous other inequalities that characterize modern life, ranging from website popularity through sexual contacts all the way into business relationships. These high concentrations of resources, information, or experience in some individuals is known in colloquial language as "winner-takes-all" or "rich-get-richer" phenomena and are more formally known as the Matthew effect.[22] Thus, the analyst needs networks that capture those results by better understanding the underlying dynamics that cause them.

Network growth may be a candidate for that. Real-life networks are not static constructs but rather grow and develop over time. Even with the same probability of obtaining a link, older nodes that have been in presence longer have also exposed themselves more to this possibility and have thus gained more links. This is why seasoned executives have more industry connections than newly inducted interns. Another explanation may have to do with preferential attachment. This term refers to the tendency of agents that prefer to connect to already highly connected agents because of a variety of reasons. In the world of research, a highly cited paper attracts the attention of scientist and thus increases the probability of them reading and citing it. This, in turn, make it even more widely cited and starts a virtuous cycle (for its author) of increasing popularity. Outside science, preferential attachment is the part and parcel of influencers and viral marketing.

The Barabási-Albert model tries to capture both drivers that can generate those dynamics.[23] It expands the random network by including the following characteristics:

- **Growth**—at each cycle a new node is added that connects to already existing nodes;
- **Preferential attachment**—the probability $\Pi(k)$ of a link of the new node to node i depends on the degree of this node i, or:

$$\Pi(k_i) = \frac{k_i}{\sum_j k_j} \tag{4.13}$$

While this mechanism is probabilistic, Eq. (4.13) implies that if a node has twice as many links, new nodes are twice as likely to connect to it. One can approximate the rate at which a node in the network acquires new edges as new nodes are added. This is given by the following equation:

$$\frac{dk_i}{dt} = m\Pi(k_i) = m\frac{k_i}{\sum_j k_j} \tag{4.14}$$

We denote with m the number of nodes that a new one links to. It follows that the existing node i has m chances to be connected to, and thus it follows that the sum of the denominator can be expressed as:

---

[22] Merton [33].

[23] Barabási and Albert [3].

$$\sum_{j=1}^{N-1} k_j = 2mt - m \tag{4.15}$$

Substituting Eq. (4.15) into Eq. (4.14) and approximating for cases when t is large, one obtains the following:

$$\frac{dk_i}{k_i} = \frac{1}{2}\frac{dt}{t} \tag{4.16}$$

To obtain the degree of a given node as a function of time, we integrate Eq. (4.16):

$$k_i(t) = m\left(\frac{t}{t_i}\right)^{\beta} \tag{4.17}$$

In this case β is the dynamical exponent of the network and has a value of ½. In short, Eq. (4.17) implies that the degree of each node increases following a sublinear power law with exponent β. Older and newer nodes thus compete for new links, and older ones have a significant first-mover advantage. Overall network degree distribution can be approximated using the following expression:

$$p(k) \approx 2m^{1/\beta}k^{-\gamma} \tag{4.18}$$

The Barabási-Albert model thus models the degree distribution of the network as a power law following the exponent γ, with $\gamma = \frac{1}{\beta} + 1 = 3$. The relationship between the degree exponent and the dynamic exponent here reveals the connections between network structure and its evolution over time. A basic Barabási-Albert network is presented in Fig. 4.8.

The key question here is how close this model resemble actual networks. It turns out that the degree distribution of many real-life networks can be neatly approximated by a power law distribution of the following form:

$$p(k) = Ck^{-\gamma} \tag{4.19}$$

Using normalization, we obtain an expression for the parameter C as well:

$$C = \frac{1}{\int_{k_{\min}}^{\infty} k^{-\gamma}dk} = (1-\gamma)k_{\min}^{\gamma-1} \tag{4.20}$$

Networks that follow the power-law distribution are called scale-free networks, or alternatively are said to have the scale-free property. The key parameter here is the exponent γ, which governs network topology. More specifically, in a scale-free network there is a large number of nodes with relatively low degrees, which is uncommon for random networks. Second, there is an excess of nodes whose degree

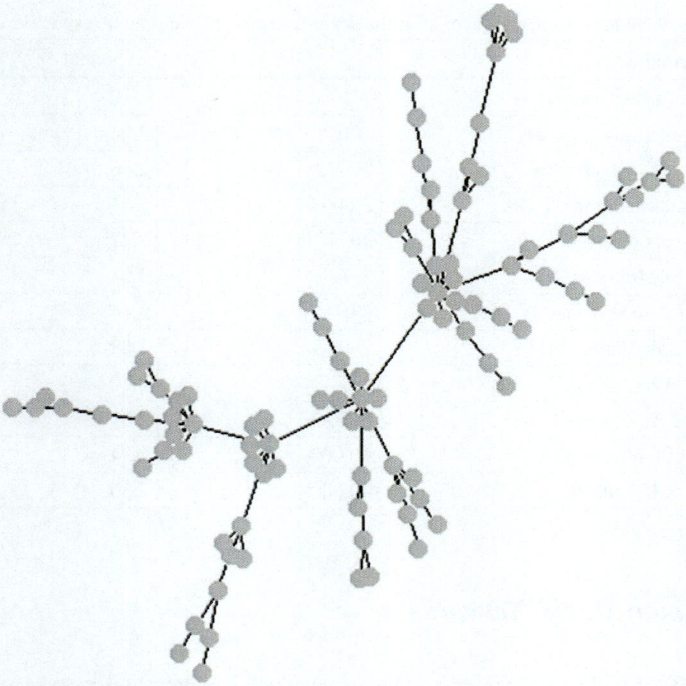

Fig. 4.8 Barabási-Albert Network with 100 Vertices and Exponent 1.2

is equal to the average expected degree; again, this is not the case in an Erdös-Renyi network. Finally, the probability of observing a high degree node is orders of magnitude greater than in a random network. Thus, in a scale-free network, highly connected components (hubs) tend to form. One may even go one step further, showing that the underlying network structure or logic, such as preferential attachment or optimization, lead to distributions that follow a power law.

It turns out that scale-free phenomena are ubiquitous and can be found in word frequencies in texts, citations of scientific papers, web hits, telephone calls, magnitude of earthquakes, intensity of wars, frequencies of family names, wealth of the richest, population of cities, and many others.[24] Representing disparate complex systems as networks, governed by a power law enable the analyst to understand the deeper order behind seemingly different phenomena and to model them with similar mathematical tools. The value of the exponent γ clearly gives idea of the rate of growth and possible inequality, and in cases of $\gamma > 1$, this spells an explosive growth trajectory that may not be sustainable, and thus reflects a possibly catastrophic risks.[25] Table 4.1 presents the exponents of a few real-life phenomena that may be useful for parameterizing risk management models—networks or otherwise.

[24] Newman [35].

[25] West [52].

Table 4.1 Selected phenomena that exhibit the scale-free property and their respective exponents

Type of phenomenon	k_{min}	Exponent, γ
Frequency of use of words	1	2.20
Number of citations to papers	100	3.04
Number of hits on web sites	1	2.40
Copies of books sold in the US	2,000,000	3.51
Telephone calls received	10	2.22
Magnitude of earthquakes	3.8	3.04
Diameter of moon craters	0.01	3.14
Intensity of solar flares	200	1.83
Intensity of wars	3	1.80
Net worth of Americans	$600 m	2.09
Frequency of family names	10,000	1.94
Population of US cities	40,000	2.30

Source Newman [35]

4.3.3 Small World Networks

In the 1960s Stanley Milgram, an American social psychologist, asked what he termed "an intelligent friend" how many intermediaries does one need to transmit a message across the United States and reach a target person at the other end of the country. The estimate stood at about a hundred persons. Milgram set out to test this and selected two target receivers and then started giving out the message to random people and followed the chains until the messages reached their destination.[26] Results were startling—the chains had a median of five intermediaries, with the distribution peaking at six. Thus, the idea of six degrees of separation was born—it is the belief that any person on the planet can be reached by hopping through at most six intermediate persons. It short, this is part of the scientific basis behind our human intuition that it is a small world that we live in.

Naturally, the exact numbers of the degrees of separation would crucially depend on how large the network is, and how well it is connected. An immense and constantly growing network such as the internet would have more than an average of six degrees of separation between two randomly selected nodes. On the other hand, a social network like Facebook that encourages connections may lead to a dramatic decrease in the degrees of separation. In fact, it is calculated that within the Facebook, there are about four degrees of separation between two randomly selected users.[27] At any rate, the very presence of such "small-world" phenomena points to an important characteristic of real networks—the average shortest path between two nodes is in

[26] See Milgram [34].

[27] Backstrom et al. [1].

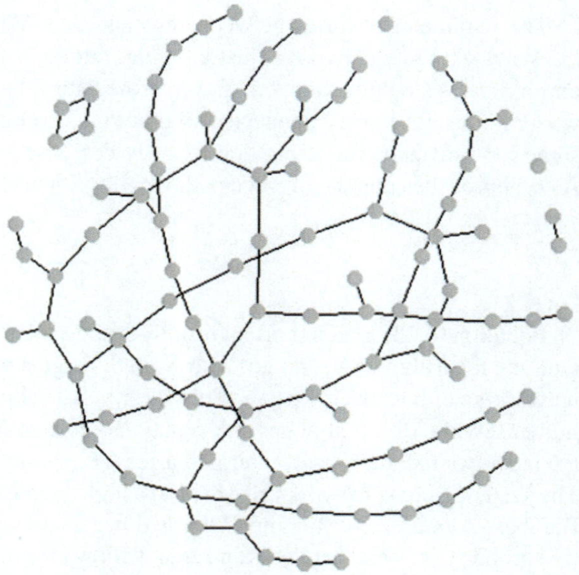

Fig. 4.9 Small-world network with 100 vertices and rewiring probability 0.25

fact significantly shorter than what should be observed in a random network, or, for that matter, in Barabási-Albert networks.

To capture the small-world property, Watts and Strogatz propose a model of a regular lattice with random rewiring that supports both the creation of links to close nodes (that dominate most real networks) but also the possibility to connect to distant nodes, which leads to shorter paths. Those, in turn, result in small-world effects (Fig. 4.9). The Watts-Strogatz algorithm aims to create a network of N nodes with a mean degree k, relying on rewiring between nodes with a probability α. It proceeds as follows:

1. Construct a regular ring lattice with N nodes, each of them connected to $k/2$ neighbors on each side (for a total of k links). There is an edge between nodes i and j if the following holds:

$$0 < |i - j| mod\left(N - 1 - \frac{k}{2}\right) \leq \frac{k}{2} \tag{4.21}$$

2. For every node, take its edge and rewire it (connect to a randomly chosen node) with probability α while avoiding self-loops and link duplication. The number of non-lattice edges k_α is thus given by the following:

$$k_\alpha = \alpha \frac{Nk}{2} \tag{4.22}$$

The parameter α varies between zero and one. When it takes a value of zero, the Watts-Strogatz network is just a regular lattice, and when it is one the network approximates a random network. For intermediate values of α one obtains a network where nodes are mostly connected to other nodes close to them but there are also some long-distance links that dramatically decrease average paths in the network. As α approaches one the average path length $\updownarrow(\alpha)$ takes the following form:

$$\updownarrow(\alpha \to 1) \approx \frac{\ln N}{\ln k} \tag{4.23}$$

Equation (4.23) is an important result—it shows that the average path length scale with the logarithm of N, and not with N. Path lengths thus do not scale linearly with network size, but much less so. This means that while those networks may grow immensely the different nodes still remain accessible from distant other nodes with a relatively small number of intermediaries (i.e. within a few degrees of separation). The Watts-Strogatz networks also exhibit a high degree of clustering that random and Barabási-Albert network do not. On the downside, they fail to reproduce the scale-free property of actual networks and do not allow for growth. Despite their limitations Watts-Strogatz networks provide important insights on network diffusion processes that may engender risk capable of quickly spreading through a set of interconnected agents.

4.3.4 Network Metrics

Networks can be aptly characterized by a number of relevant metrics but for the purposes of risk management it is often useful to focus on the characteristics of individual nodes and their overall context. This is due to a simple reason—it is often that the individual agent (vertex) is the source of risk and understanding the node's characteristics and position in the overall network allows the analyst to follow processes such as diffusion or contagion stemming from it. It is precisely the ability to model those flows that make network science useful for risk analysis. The simplest metric of connectivity in the network is the degree of a given node. The in-degree correspond to the number of incoming links to the node. Thus, if a given network is described by adjacency matrix **A**, then the total sum of the links to node i coming from a number n of other nodes j is its in-degree k_i:

$$k_i = \sum_{i=1}^{n} A_{ij} \tag{4.24}$$

Similarly, the out-degree k_o describes all the links going from the specific nodes to all others it connects to, or:

$$k_o = \sum_{j=1}^{n} A_{ij} \tag{4.25}$$

The degree of a node is thus the sum of its in-degree and its out-degree. Highly connected nodes (i.e. network hubs) that can be conducive to spreading risk across the networks have a higher degree. The overall network topology can also be characterized by the network's average degree. Higher value of this metric indicate dense network, whereas lower value indicate much sparser graphs. It is often of interest how given parts of the network (or sub-graphs) are connected. Intuitively, those connecting nodes are the brokers in the network, and their removal may lead to disintegration. Such an important vertex is characterized by the fact that many of the shortest paths from one node to another pass through it. A measure of how many of those paths go through that special node is the betweenness centrality of the node. If one denotes the shortest paths connecting jk as g_{jk}, and the path that node i lies on as $g_{jk}(i)$, then its betweenness $C_B(i)$ is given by the following:

$$C_B(i) = \sum_{j<k} \frac{g_{jk}(i)}{g_{jk}} \tag{4.26}$$

Betweenness centrality is important in terms of identifying bottlenecks, single points of failure or points through which the flow must pass. However, it may be possible for a node to efficiently participate in the flow even if it is not on many of the shortest paths, but rather it is close enough to those. Such nodes can be identified by measuring their closeness centrality—the higher this score, the smaller the distance to all other nodes. If the distance between node i and a set of nodes j is denoted as $d(i, j)$, then the closeness centrality $C_C(i)$ takes the following form:

$$C_C(i) = \left[\sum_{j=1}^{N} d(i, j) \right]^{-1} \tag{4.27}$$

Nodes that are characterized by high closeness centrality can easily spread innovation, information or contagion to other nodes and can thus have an outsized influence on the network. Both the degree distribution and node centrality metrics aim to identify which are the important and influential nodes that can be the targeted—either as an amplifier (e.g. when spreading a message) or as a source of problems (e.g. in the case of pandemics). A final important characteristic would be the context in which those key vertices are situated. The degree of a given node, or its centrality, provides little information about how the nodes around it are connected (interact) with each other. This can be measured by the degree of clustering, as captured by the local clustering coefficient C_i. It varies between zero which indicates that there are no links between i's neighbors, and one which indicates that they are fully connected. The local clustering coefficient for an undirected network is given by the following expression (with $\mathbf{A}$ again denoting the network adjacency matrix):

$$C_i = \frac{1}{k_i(k_i - 1)} \sum_{j,k} A_{ij} A_{jk} A_{ki} \tag{4.28}$$

In the case of random network, C_i is given by the following:

$$C_i = \frac{2\overline{k_i}}{k_i(k_i - 1)} = p = \frac{\overline{k}}{N} \tag{4.29}$$

This coefficient provides an idea of how dense a given node's neighborhood is, and together with the node's specifics (e.g. high degree or centrality) enables the analyst to fully characterize the local impact of a risk arising in node *i*. The local clustering coefficient can be extended to a global one, and the assiduous analyst will find other useful metrics that characterize the network as a whole, as well as its individual elements. The key message, however, aptly carries across the diversity of measures. It is that the topology of the network, and the specifics of each vertex have crucial implications for risk management, whereby highly connected and central nodes have a significantly disproportionate effects on observed outcomes—both positive, as well as negative.

4.3.5 Network Robustness and Resilience

The network topology allows the analyst not merely to understand diffusion processes across networks but also to conceptualize the different modes of network failure and their ensuing effects. Network robustness and resilience are focal topics in studies of all types of networks—ranging from technical through biological to socioeconomic ones.[28] Moving away from analyzing the risk of a single actor, organization, position, or portfolio, and into a systemic and holistic approach is a much-needed stride for ensuring the robustness and resilience of complex systems.[29] This is an equally large stride for risk management. While it is imperative from the standpoint of the ever-increasing complexity and risk exposure we are faced with, it is also bedeviled by the relatively limited data availability of interconnections (links) in the fields of business, economics, and society.

It is instructive to make a distinction between robustness and resilience.[30] Resilience shows the tolerance of the network to a large change (perturbation) without shifting to an alternative steady state. Essentially, it shows the ability of the systems to absorb shocks over time. Robustness is a more narrow concept—it shows the static ability of the system to absorb a blow without ceasing to function, as a number of links are no longer operational or in existence. While resilience is crucial over time, usually risk management tends to focus on worst case outcomes that generally occur

---

28 Liu et al. [30].

29 Smith et al. [42].

30 For definitions, refer to Liu et al. [30].

after the realization of the risk but before the system has adapted itself to it. One may thus find insight in understanding the robustness of networks.

This can be studied formally by investigating what happens to different types of network topologies as a given fraction f of their nodes are removed. Unsurprisingly, results vary widely depending on the type of network and the ensuing effects as nodes are removed.[31] In the case of failure, one assumes that f random nodes are no longer in existence or stop functioning. Within a purely random Erdös-Renyi network there is a critical number of nodes f_c that when removed lead to the disintegration of the network's giant component. In short, once the critical threshold is reached, the random network loses connectivity and stops functioning as it used to. On the other hand, in the case of scale-free networks one does not observe this effect. It is only as f_c approaches one that the network connectivity fails. The intuition behind this is straightforward—in a scale-free network there are many nodes with few connections and very few hubs with many links. Random removal has a much higher chance of hitting a weakly connected node (there are many) than a strongly connected one (there are few).

Obviously, this intuition only works as we observe random failure. On the other hand, it may very well be the case that the network is not subjected to a random failure but to a concerted attack by an actor who has a good understanding of the network topology, and the ability and desire to bring the network down by targeting specific vulnerable nodes. It turns out that especially in scale-free networks the critical point at f_c reemerges very quickly and has a remarkably low value. One can easily break a power-law network by removing a few of the highly connected hubs, and in case of attacks scale-free networks are more vulnerable than random networks.

Networks can clearly be exposed to both types of disturbances—failures and attacks—and their overall robustness depends on their ability to withstand well both. However, those may sometimes come in conflict and thus when designing for robustness may have to take the tradeoff into account. For example, a perfect hub-and-spokes network with one central hub is very safe in regard to random failure—the chance of the central hub actually failing is very small, and the failure of any other hub does not destroy the network. In contrast, a targeted attack on this central hub will break it for good—it is the single point of failure. Thus, one needs to maximize robustness for a total critical threshold f_c^{tot}, , that includes the thresholds for random (f_c^r) and non-random disturbances (f_c^a), or:

$$f_c^{tot} = f_c^r + f_c^a \tag{4.30}$$

It turns out that this is best achieved by the bimodal distribution of the following form[32]:

$$p_k = (1 - r)\delta(k - k_{\min}) + r\delta(k - k_{\max}) \tag{4.31}$$

[31] Barabási [2].

[32] Paul et al. [37], Tanizawa et al. [47].

In Eq. (4.31) r denotes a fraction of the nodes that have a degree of $k_{\max}$, while the rest $(1 - r)$ have a degree $k_{\min}$, δ is a tuning parameter. Naturally, real-life networks do not follow this distribution for a variety of technological, historical, behavioral, economic, or political reasons. Still, Eq. (4.31) presents a useful benchmark for the distribution of a truly robust network.

A final consideration with networks is the issue of cascading failures. Those are most spectacularly observed in electric grids as problems in one node leads to the electricity being rerouted through to other nodes. Some of them may be unable to take the increased load, thus failing themselves, until this domino effect leads to massive blackouts. Similar dynamics can be observed in Denial of Service (DoS) attacks, and financial crises. For a cascading failure to occur, the network must have the following three characteristics:

- There is a flow over the network (information, electric current, financial flows);
- Each node has a capacity and a local breakdown rule, determining whether it functions as intended, or not;
- There is a mechanism for redistribution over the network (of traffic, information, money).

The specific types of cascading behavior will depend upon the network topology, and the behavior of individual nodes within the network. This brings about a broader conclusion about modeling network topologies—while models such as the ones presented here give key insights about the behavior of real networks, precise behavior and effects for the purposes of risk management can only be obtained by leveraging actual network data and modeling and interpreting it within the context of extensive subject matter knowledge.

4.4 Network Construction

The key insight stemming from network theory is that the degree of interconnectedness between agents, actors, processes, and events may have critical implications for the flow of risk across the network and for the quantitative impact of uncertain negative occurrences. Moreover, network analysis enables the understanding of events that go beyond the single node and can have catastrophic effects at the systemic level, such as cascading failures. The actual application of network knowledge for the need of research or practical endeavors crucially depends on data type, quality and availability. Those three characteristics serve as useful constraints that inform and guide the analyst as to the scope of network science applications. One can thus review two standard cases—the case where is present for both nodes and their links, and the case where data on the nodes alone is present. The third case—where data on neither exists—is relegated to the next chapter.

4.4.1 Network Data and Visualization

Data for network visualization may be obtained from different sources. Social Network Analysis (SNA) usually focuses mostly on networks with human participants. Datasets thus stem from online social networks whereby agents and their interactions are captured, from organizations which map the connection between stakeholders, or from primary data collection as is the case with the archetypal research problem of intertwining corporate power. Such datasets are difficult to obtain as they may represent proprietary internal data that companies may view as a source of competitive advantage (as is the case with online social networks) or are highly sensitive and individuals may be unwilling to share freely (as is the cases with elite connections). While the researcher may always opt for a primary data collection process this naturally limits the extent of the network and only helps advance the solution to simple problems that may benefit from a limited network of tens or hundreds of individuals.[33] Networks of thousands or millions are practically impossible to collate in an entirely manual process. Data availability is thus a natural constraint to fully leveraging SNA[34] for risk management purposes.

An application of SNA to macroprudential policy has an insightful strand of network literature, whereby agents are not physical persons but rather legal entities (e.g. financial institutions). Here, the individual nodes tend to interact with others, and a usual way to operationalize the links is the flow of funds or counterparty exposure.[35] The key problem here stems from the availability of high-quality public data on mutual exposures between financial institutions and their development over time. It is not common practice for regulators to require the publication of such data, and financial institutions are naturally reticent to disclose the most intimate details of their balance sheets. So, in this case the analyst has access the plentiful data about the nodes but limited data about the edges. Researchers have tried to circumvent this thorny problem by estimating a proxy for the relationship between financial institutions such as their effects on counterparty risk metrics (e.g. the VaR).[36] While insightful, this approach gives a glimpse into overall structure and may lead to biased quantitative estimates. As of now, a fully satisfactory solution is yet to be found.

Another large groups of network science applications to economics has been in the field of international trade.[37] Because of its specifics and for historical reasons, international trade data is very detailed and generally publicly available. This makes the network construction a straightforward task with the trading partners being the network nodes, and the imports and exports of the good or service of interest—the edges. As the network topology accurately reflects legal trade, it is particularly useful to study the interconnections and interdependence between different countries, and thus their likelihood of engaging in specific activities (such as trade or war). More

---

33 Helms [22].

34 Srivastava et al. [43].

35 Such is the model by Gai and Kapadia [18], presented in this chapter.

36 Hautsch et al. [21].

37 DeBenedictis et al. [15], Chaney [10], de Andrade and Rego [14].

poignantly, the network structure of international trade allows the analyst to model the transmission of economic shocks on the global stage and estimate the effects on individual trading partners as well as the spread of such a contagion. From a risk management perspective this is useful in two main directions. First, it enables one to model external effects on specific industries, thus capturing a component of specific and diversifiable risk. Second, it provides a proxy for exogenous macroeconomic shocks that may impact growth, and thus reflects a part of potentially non-diversifiable market risk.

Apart from purely economic and business applications, network analysis has focused its study on critical infrastructure networks such as utilities (water, power grid) or communications (road, air, rail), the internet, mobile networks, brain networks, biological and ecological networks, as well as many other network-related phenomena.[38] While providing enlightening results, not all network phenomena can directly inform risk management. On the other hand, data on the network structure of relevant phenomena such as internet connectivity, individual proximity to different points, or ecosystem tipping points are rarely publicly available and may sometimes be considered sensitive by the respective stakeholders. In short, current state of network data availability presents a constraint on the full-scale rollout of network science approaches for all risk management problems.

The best-case scenario for the ardent network analyst is when complete information of the network topology is present. In such cases data on nodes and links is available and is of sufficient quality. The usual way to store is through an adjacency matrix but alternatives such as the adjacency list are also useful and common. Numerous software utilities allow the processing and visualization of such data whereby agents, processes or events are the nodes, and their interactions—the links. Peculiarities of the nodes may be encoded visually—e.g., the relative importance of the node may be given by a change in its size, and its class—by some coloring. Likewise, edges may be of different types with width corresponding to intensity and the type of line—to different types of interactions. In case only part of this data is available, the analyst may still reconstruct the network by taking recourse to advanced statistical techniques. One of them is the Bayesian Belief Network.

4.4.2 Bayesian Belief Networks

It is the common case that the analyst has point data on different phenomena but only conceptual understanding of the links between them. An apt example would be revenue analysis—there may be enterprise data on sales and inventories via transaction systems, on process efficiency through a process-aware system, online user behavior via digital channels, and possibly customer satisfaction ratings obtained through surveys. While it is plausible that links exist between those nodes of the firm network, the specific causal links and their quantitative impact remain unclear.

[38] Brugere et al. [7], Liu et al. [30].

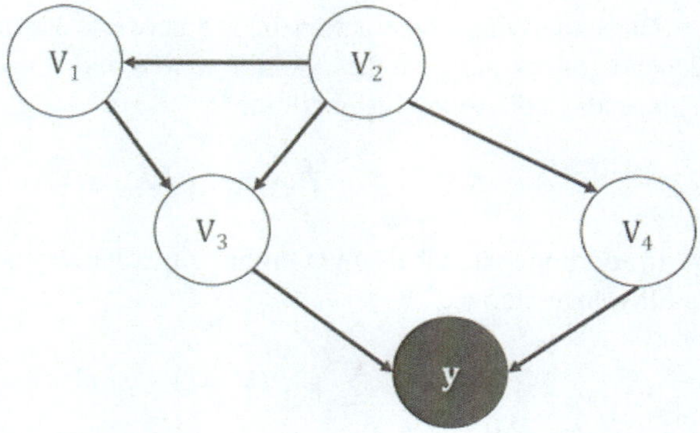

Fig. 4.10 A schematic Bayesian network

It is thus impossible to conduct a rigorous risk analysis with this data alone. One can thus use statistical technique to reconstruct a network of causal links between the variables of interest and measure their relative effects using a Bayesian Belief Network (BBN).

The Bayesian network is a model that takes a probabilistic view of the world. Moving beyond traditional deterministic causal links, it recognizes that effects of one driver on an end result may be uncertain and are thus best modeled using a conditional probability distribution. Let us denote the probability of a given event A occurring with *P(A)*, and that of an event *B*—with *P(B)*. The well-known Bayesian theorem relates the probability of A occurring, given that B has occurred, or its conditional probability *P(A|B)* as follows:

$$P(A|B) = \frac{P(A \cap B)}{P(B)} \tag{4.32}$$

Thus, the analyst may estimate an unknown probability (e.g., increase in revenue) given a set of known ones. Building upon that, the Bayesian network creates a whole set of interconnected events (network nodes) that each has its own specific probability distribution. The edges of this network are the marginal contributions of given variables to the probability of realization.[39] In short, the Bayesian Belief Network constructs a visualization of variable (events, processes, etc.) of interest that present their causal links in the form of a direct acyclic graph (DAG). Such a graph is presented visually in Fig. 4.10. In this case variable V_1 affects V_3, V_2 affects V_1, V_3, V_4, and the target variable y is affected by V_3 and V_4. The transmission of risk within the system is straightforward to conceptualize: once a failure happen in V_1 or V_2 it cascades through V_3 to the target y. On the other hand, problems with V_4 directly affect the target. What is more, this model enables the analyst to conduct deeper investigation of effects, as the BBN constructs a local conditional probability distribution for every node.

[39] For more details, refer to Cowell [12], Lauritzen [29], Korb and Nicholson [27].

The underlying construction of the network assumes that any given node V_i depends on its parent nodes alone ($pa(V_i)$) and no other variable in the system. This leads to the following condition[40]:

$$P(V_i) = P(V_i|pa(V_i)) \tag{4.33}$$

In addition to that the sum of probabilities at every node is one, and thus the event is fully characterized[41]:

$$\sum_{V^*} P\big(V_i = V^*|pa(V_i)\big) = 1 \tag{4.34}$$

Apart from the individual conditional probabilities, the network allows the computation of an aggregated probability distribution $P(V)$ that is equal to:

$$P(V) = \prod_{V_i \in V} P(V_i|pa(V_i)) \tag{4.35}$$

Here one should note that the Bayesian network is not a deterministic model—there may be slight variations depending on the computation algorithm chosen or the criterion for ending the optimization procedure. While the increase in the sample and quality of the sample should minimize those effects, the analyst may want to gain a better understanding of how the network is computed, so that results may be correctly interpreted.

The Bayesian network is iteratively learnt given a sample of data. The standard approach to this typically consists of three phases.[42] Initially the algorithm learns the structure of the network (i.e., the presence of edges between different nodes), and then at a second stage determines the direction of those edges. The latter implies causality links. Network construction is then repeated over numerous iterations and in the third stage, the optimal network is chosen. An information criterion, such as the Bayesian (BIC) one, is used for this determination.

The structure can be learnt using two basic approaches—either through constraint-based or through score-based algorithms. Constraint-based algorithms use the Markov property of networks to formulate a set of constraints; the network is then fitted within those. Score-based algorithms work by assigning a score to each network constructed and then selecting the "winner" given this score. This may either be a variant of a likelihood or log-likelihood score, or an appropriate information criterion (AIC, cAIC, BIC, etc.). While there are different computation approaches in the case of discrete and continuous data, the network interpretation remains essentially the same and the choice of algorithm bears little effect on the final application for risk management purposes.

40 Bøttcher and Dethlefsen [6], Højsgaard [23], Wang et al. [49].

41 Ibid.

42 Margaritis [31], Daly and Shen [13], Scutari [40], Wang et al. [49].

Table 4.2 Contributions to conditional density of target variables in next period

Variable	Unit	Employment		Turnover	
		J62	J63	J62	J63
Gross Domestic Product, GDP	CLI	0	0	0	0.667
Household Consumption (C_H)	% GDP	0	0	0	2.337
Export of Goods (X_G)	% GDP	−0.222	0	0	1.014
Import of Goods (M_G)	% GDP	0	0	0	−0.702
Export of Services (X_S)	% GDP	0	0	0.153	0
Import of Services (M_S)	% GDP	0	0	0	0.457
Interest rate (r)	%	0	−10.386	0	−4.547
Inflation (HICP)	CLI	2.238	0	0	0.986
Labor costs in J60 (C_L)	CLI	0	0	0	0.756
Compensation of Employees (W)	CLI	0	0	0.741	1.426
Unemployment	%	−1.557	1.107	0	0
Sample Size		591	591	1283	1283

Note Only variables with edges to target variables included in table

4.5 Network Sources of Risk

The overall network structure within which an industry or a company operates has crucial repercussions for risk management. On the one hand it allows better identification of risks, and thus improved analysis and mitigation. On the other hand, dynamics in proximate variables may be used as a leading indicator for upcoming problems. Industry and firm-level modeling is further eased by the large availability of public data that can be complemented with private sources so that the environment is well-described. The plethora of publicly available data allows for a particularly detailed analysis of non-diversifiable risk, as it stems from global phenomena such as macroeconomic, social, or demographic dynamics. This is also the focus of this section—a particular investigation of the drivers of systemic risk for the digital industry.

4.5.1 Overall Dynamics of the Digital Sectors

When analyzing the peculiarities of the digital industry, the analyst is often confounded at the very first definitional step as to what exactly constitutes the digital industry. Indeed, this is a question that has inspired much research.[43] While it is clear that IT-related sectors form the backbone of the industry, many traditional enterprises

[43] Barefoot et al. [4], Watanabe et al. [50].

have extensively leverages information and communication technologies, thus effectively blurring the boundaries between them and the IT sectors. This has resulted in a plethora of names that are meant to emphasize this—"fintech" that stands for the merger between finance and technology, "edtech"—for the one that pertains to education, "healthtech"—for digital healthcare, "regtech"—for compliance solutions, and so on. To avoid confusion on the matter, one may wish to focus on the core providers of IT products and services as laid down in national and international sectoral classifications. More particularly, the European Statistical Service leverages the Statistical Classification of Economic Activities in the European Community (commonly known as NACE) ver. 2 to classify different economic activity, and most core ICT production falls under code J60. Excluding purely telecommunication services (code J61), one is left with the following two categories:

- Code J62: Computer programming, consultancy, and related activities;
- Code J63: Information service activities.

While imperfect, those two clearly defined sectors are able to capture the crux of the information economy. The former, J62, proxies the production of new digital solutions. The latter, J63, is focused on service provision that is directed toward intermediate and final individual and business consumers.

The explosive growth of the digital sectors can be clearly seen in Figs. 4.11 and 4.12. We leverage Chain Linked Indices (CLI) of sectoral turnover, employment and the growth of the overall economy, with all time series being seasonally and calendar adjusted. In Fig. 4.11 all data is presented as a CLI with base year 2000 and one can clearly observe the explosive growth of computer programming and related consultancy services that has wildly outpaced the aggregate growth of the overall economy. This data signals at structural shifts that point to the increasing importance of the digital sector and thus—the increased potential of IT to generate systemic level risks—both on the upside, and on the downside.

The situation is quite similar in the other sector of interest. Growth in information services, which also includes data processing, has been similarly spectacular (Fig. 4.12). The figure presents visually data for four selected European countries but those are quite representative of the whole sample. A quick note may be in order here. There seems to be very different growth in terms of information sector activities. For the period of two decades developed countries such as Germany, France, and the United Kingdom have managed to triple or quadruple the IT sector while developing ones such as Bulgaria have seen a 60-fold increase. This is due to two major trends. First, this is a base-level effect—starting from a lower base growth in relative terms seems proportionately larger. Second, and more importantly, there is a clear trend that highly active clusters of digital services develop through some parts of Europe. This is largely driven by the logic of exponential growth of successful IT businesses and a reliably decreasing marginal costs of production. For the purposes of risk management we note that this may engender the need to model ICT-related phenomena using power law distributions.

Finally, one of the recurring themes regarding the digital sectors is the so-called "productivity paradox". It pertains to the fact that while the IT sectors have

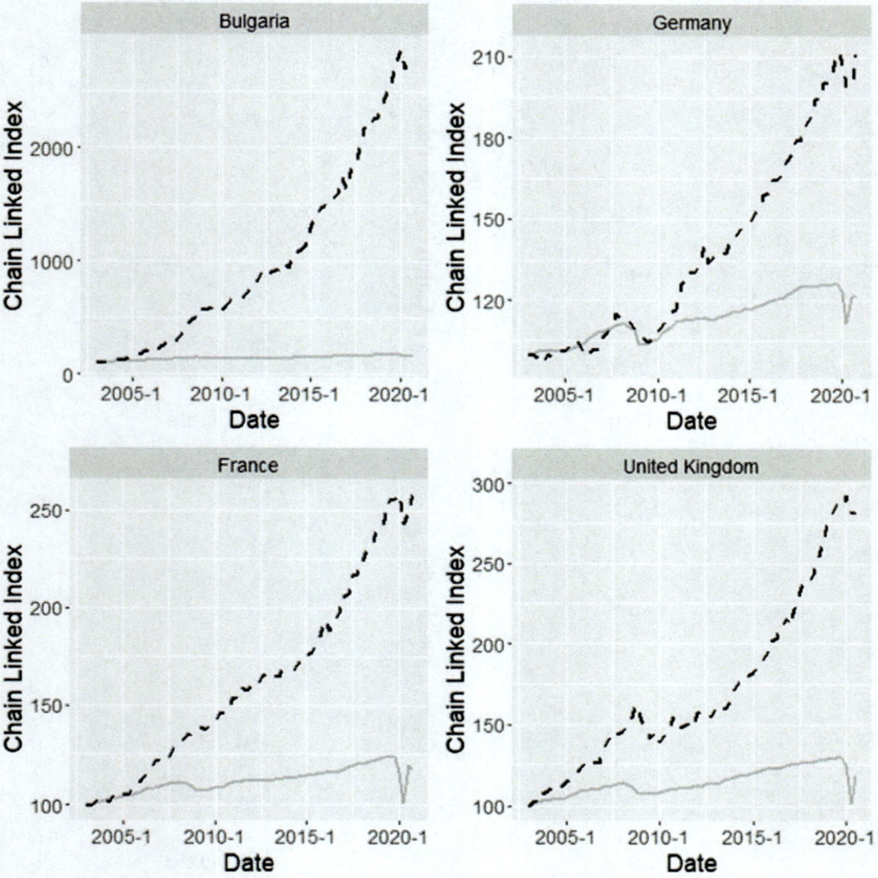

Fig. 4.11 Growth in J62 Turnover (dashed black line) and Overall Economy (solid gray line), Chain Linked Index rebased to 2000, selected European countries

grown tremendously and rapidly accelerated the digital transformation of the overall economy, this has not resulted in significant measurable increases in productivity.[44] Some authors argue that this is because digital solution need an incubation period ("installation phase") before they are fully absorbed into business processes and economically useful.[45] The two sectors under study here (J62 and J63) may give hints as to the viability of this hypothesis.

Figure 4.13 shows the dynamics of turnover and employment in those over a period of two decades (2000–2020). Data clearly shows that initially turnover was almost perfectly tracking employment, thus showing relatively small productivity effects. Essentially, companies were growing by hiring more people. However, by the end of the period turnover growth started significantly outpacing the growth of the

[44] Van Ark [48].

[45] Ibid.

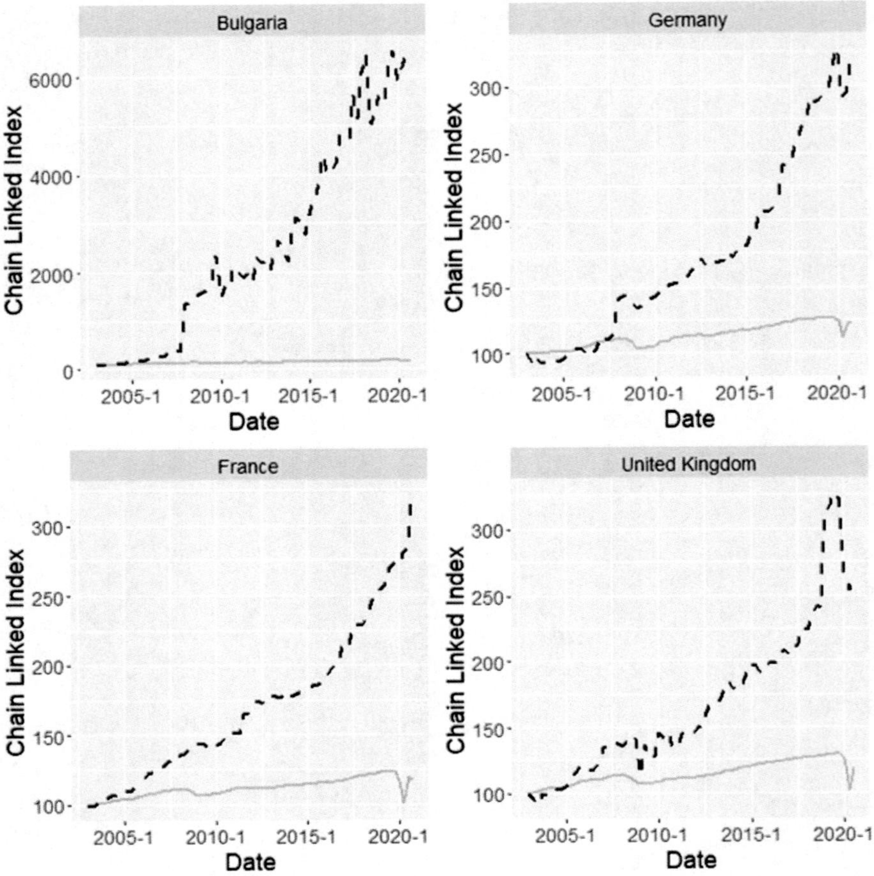

Fig. 4.12 Growth in J63 Turnover (dashed black line) and Overall Economy (solid gray line), Chain Linked Index rebased to 2000, selected European countries

workforce. This hints at increases in productivity that are only visible after some time, probably due to operational lags and learning curve effects. This is likely to be seen throughout the economy, as more traditional sectors become digitally transformed. After understanding overall dynamics of the digital sectors, it is crucial to also see what drives them and what is the quantitative magnitude of those influences.

4.5.2 Network Sources of Risk for the Digital Sectors

We aim to better understand the sources of systemic risk that affect the digital sectors. To this end the analyst may create a network of interactions and investigate which systemic (macroeconomic and sectoral) variables have a consistent influence on the

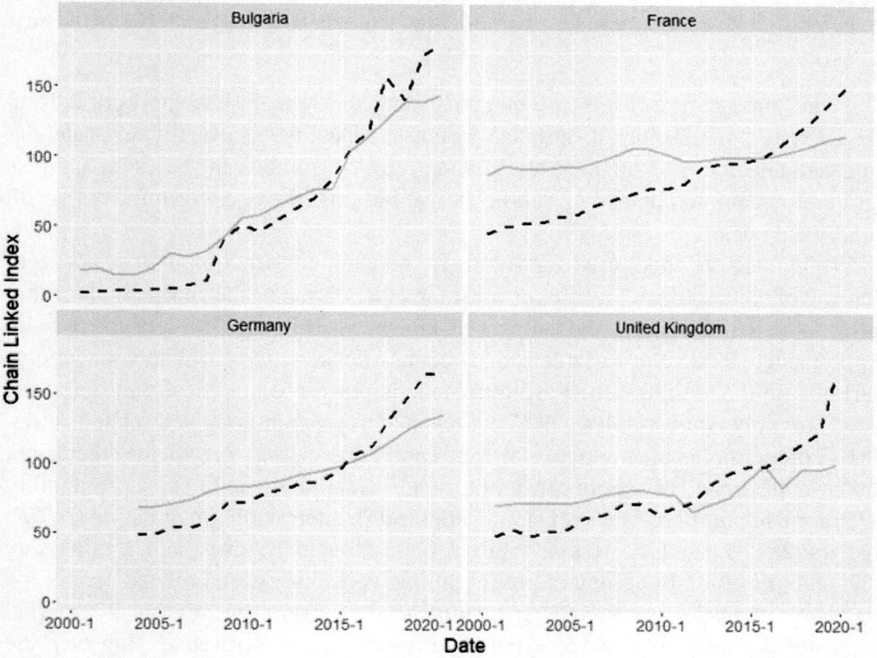

Fig. 4.13 Growth in J63 Turnover (dashed black line) and J63 Employment (solid gray line), Chain Linked Index rebased to 2000, selected European countries

development of the ICT industries. The focus on non-diversifiable systemic risk is important to investors and decision-makers as it cannot be hedged away by investing in an alternative digital solutions provider or relying on another IT vendor. Additionally, a large amount of public data is available that allows for the reliable network construction using a Bayesian approach.

The data used for this exercise comes from the European Statistical Service (Eurostat) and include three major types of data:

- **National Accounts Data**—GDP as Chain Linked Index (CLI), export and import of goods (% of GDP), export and import of services (% of GDP), household and government consumption (% of GDP), investment/gross capital formation (% of GDP), subsidies (% of GDP), and operating surplus and mixed income (% GDP) as a proxy of profitability.
- **Price and Interest Data**—a variety of pricing indicators such as the inflation as measured by the Harmonized Index of Consumer Prices, HICP (in %), labor costs in the overall economy—the compensation of employees (% of GDP), labor costs in the digital part of the economy—sector J62 (CLI), and the interest rate as a proxy for the cost of capital (in %).
- **Sectoral Target Variables**—as target variables we investigate two key indicators of sectoral health—employment (CLI) and turnover (CLI). Taking into account that dynamic effects in the economic system usually need some time, the focus

here is not on contemporaneous employment and turnover but on those in the next period.

Those time series consist of quarterly data that encompass the period from Q1.2000 to Q4.2020 for 28 European countries. Data are seasonally and calendar adjusted, thus focusing on their trends. After removing all missing cases the samples for employment end up at 591 valid observations, and the ones for turnover—with 1283 valid ones.

While there are alternative algorithms for network construction, it does not seem that any of them intrinsically outperforms the others[46] and so the choice of algorithm falls on the Hill-Climbing one that is well-known, tractable, and widely understood. Applying it to the J62 turnover data, one obtains a dense network of interconnections. Part of it in the vicinity of the turnover node is presented graphically in Fig. 4.14. There are only two variables that affect revenue in programming and related activities—compensation of employees (W) and export of services (X_S). On the other hand, this revenue drives household consumption (C_H), labor costs in IT (C_L), profitability (OS or π), import of services (M_S), subsidies (*Sub*), interest rate (r), inflation (*HICP*) and growth (Y). The last two in turn affect, quite expectedly, investment (I). In a way J62—the provision of computer programming, consultancy, and related services—acts like a typical supply sector. It is influenced by overall wages and its external demand (as proxied by service export) are major drivers of revenue. However, the turnover then spills into consumption (such as household consumption, imports, subsidies) and feeds into macroeconomic dynamics affecting growth and prices. The supply analogy is even clearer as one considers that a large part of the output of J62 is either directed to businesses or represents the digital infrastructure (an analogy to physical capital) on which solutions, products, and services run.

The situation is very different with the Bayesian network that includes J63 revenue as a target variable. Again, using the Hill-Climbing algorithm, the network is constructed and partly visualized in Fig. 4.15 with the focus being on the vicinity of the turnover node. The revenue in information service activities is affected by a wider number of drivers. It is influenced by wages (W), but also imports of goods and services (M_G and M_S), export of goods (X_G), household consumption (C_H), IT labor costs (C_L *J60*), GDP (Y), interest rate (r), and prices (*HICP*). It, however, does not influence any other nodes in the network. Thus, one can say that the information service activities function roughly as a consumption sector—it is dependent on disposable incomes (i.e., growth and wages) and prices, and susceptible to substitutes and complements from the external sector. It has limited downstream influence, as well.

Obviously, the characterization of J62 as a production sector and J63—as a consumption one is rough and imprecise. There is significant diversity of IT providers in both of them and this generalization does not do justice to all. As a coarse-grained description it may still be useful as the analyst navigates the murky waters of systemic risk. The visual representation of networks is a crucial initial step for understanding

[46] Scutari et al. (2018).

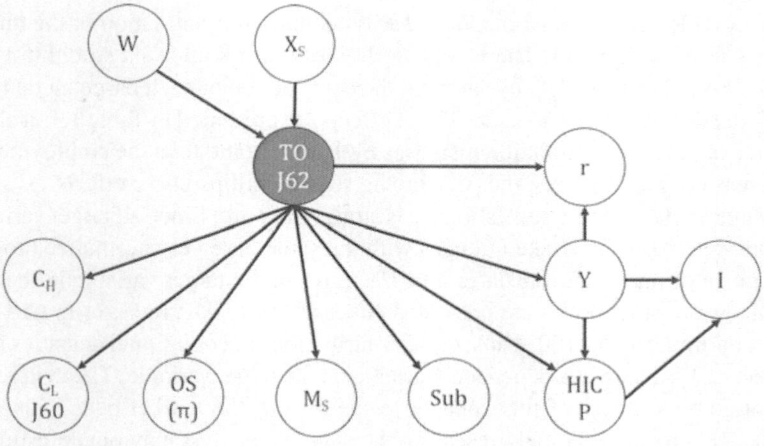

Fig. 4.14 Part of the Bayesian belief network for total turnover in sector J62: computer programming, consultancy, and related activities

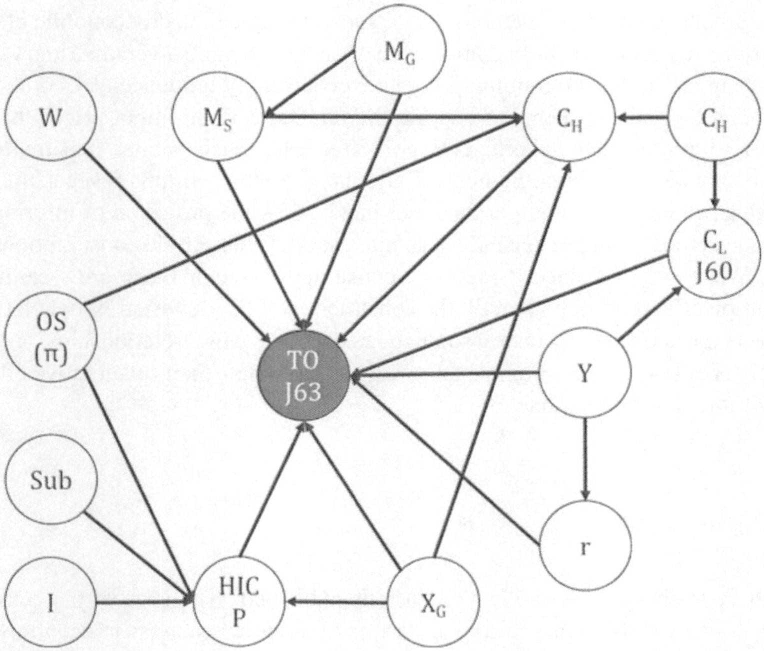

Fig. 4.15 Part of the Bayesian belief network for total turnover in sector J63: information service activities

domain complexity. The next one necessarily includes the estimation of the quantitative effect of each driver. The Bayesian Belief Network allows a straightforward calculation of driver strength by showing the contribution of each parent node to the conditional density of a child node. Those effects are presented in Table 1.1. It shows that, for example, as the inflation increases by 1 basis point, then the employment in J62 increases by 2.24 points, thus displaying strong Phillips-curve effects.

The interpretations of those estimates is straightforward. Since all target variables are represented as chain-linked indices with the same base year normalized to 100, then a move of one point translates into 1% move in the target variable in the base year. Likewise, other nodes are presented either in % of GDP, in %, or in CLI with base year normalized to 100. Thus, those contributions to conditional densities show the effect of 1% move in a dependent variable on the target variable. The avid reader may note the closeness of this concept to the idea of the market beta. This is no coincidence—by appropriately rescaling the data, the analyst may obtain estimates of percentage effects, or betas,[47] that show how sensitive a target variable is in respect to a parent node.

Employment in J62 and J63 is very sensitive to prices (inflation/interest) rate and the overall economy-wide unemployment, showing typical macroeconomic effects. On the other hand, effects on revenue (turnover) are much more diversified and varied. Turnover in computer programming is relatively strongly influenced by compensation of employees that seems to capture the effects of rising input prices that are indicative of relative labor shortage. Export effects are much smaller, presumably as they are spread over a large number of alternative sectors. In this respect J62 also acts like a production sector. On the other hand, J63—the provision of information services—is most strongly dependent on interest rates, household consumption, and wages. We argue that since it acts as a consumption sector those are wealth and substitution effects in play. Overall, the construction of the Bayesian Belief Network empowers the analyst to recreate a sophisticated web of causal relationships, uncover the sources of risk within a complex network, and estimate their quantitative effects, thus forming a rigorous model of risk.

4.6 Conclusion

In 1996 Spanish sociologist Manuel Castells published "The Rise of the Network Society"—the first book in a trilogy describing the global changes to economy and society engendered by increased connectivity, unequal diffusion of technology, and the ensuing transformation of work and life.[48] It outlined key changes in the modern world and gave insights into the future. The book was prescient, instantly popular and very provocative. Much was debated, and about half of it was later redacted, with Castells dropping more than 40% of the contents of the original book. Still,

---

[47] In economics this is sometimes called "elasticity".

[48] Castells [9].

it gave rise to a new thinking about the social-economic system as a network of interconnected agents, enabled by the digital transformation of productive activity and leisure. Despite the huge efficiency gains, such networks rapidly expand the risk exposure of a society, organization or individual—interconnectedness means more sources of failure, intense interaction means more possible issues, technology diffusion means more inequality, and the compression of time and space raises the specter of traditional thorny questions such as identity and agency.

In short, the expansion of networks also means that risk may have a network structure. This is predicated on the fact that many economic interactions are now pushed away from the traditional hierarchy of the organization and into a complex constellation of producers, their vendors, strategic partners, competitors, consumers, regulators, opinion-leaders, and so on. The modularization of digital solutions introduces further technogenic risk, and the flow of information throughout the network leads to unexpected emergent effects.

This chapter has gone through state-of-the-art network thinking in an attempt to formalize this insight. Starting from the simple premise that a network is merely a graph composed of nodes and edges, one can analyze how a random network would develop and react to disturbance. In practice, networks are more complicated than that and phenomena such as preferential attachment and small-world effects dominate their structure. In real-life networks nodes are people, organizations, processes or events and a vital payload of information or value (monetary or intangible) flows through the numerous edges connecting those vertices. Understanding the network structure may be crucial to understanding a wide range of high-risk phenomena ranging from complex financial fraud as in the case of Bernie Madoff through cascading financial failures. Even simple models can help elucidate the risk exposure.

Understanding network topology leads to the question of network resilience. While traditional applications have focused on things such as electricity grids, transportation networks or air routes, social scientists tend to focus on networks of money, power, and information. Cascading failures occur when a given node has a local breakdown rule which is dependent on adjacent ones and the failure of one vertex causes the failure of others. The archetypal example is a financial meltdown where the default of a financial institution leads to the default of its counterparties. Understanding the network topology allows for modeling the exposure and investigating the risk of such cascading failures. While it may be theoretically straightforward to devise a network with higher resilience, in practice many social and economic networks are constructed over time and within different constraints and thus deviate from the perfect model.

Social Network Analysis has gained particular prominence over the past two decades in conjunction with the rapid growth of social networks. The nodes in those networks are people and the arcs are their interactions. The insight that context and interaction crucially shape behavior means that agency and action may be influenced by other network nodes. Striking examples include voting in elections and social unrest. The information flows in those digital networks can engender real-life risks. The construction of SNA networks calls for the availability of large and detailed datasets that are rarely open to the public. Instead, they usually remain proprietary and are used by their respective owners to gain competitive advantage through advanced analytics. The risk analyst is likely to leverage SNA only as they possess the necessary internal data.

Often, data is available on the nodes of the network itself but not on their interactions. A possible approach here would be to reconstruct the network by observing node dynamics and estimating the conditional probability of a node changing given a change in another one. The analyst may use a Bayesian Belief Network to this end, whereby one obtains a probabilistic view of a complex system but is also able to pinpoint the links between different variables of interest. The strength of the edges can be estimated by calculating the contributions to the node conditional density. With the right rescaling and normalization of data, the risk analyst can obtain betas for the sensitivity of the target variable to its parent nodes. While the statistical algorithm is different, this resulting beta estimate is well in line with known and useful risk metrics and can be seamlessly included in overall risk analysis efforts. Such an approach is particularly useful when large-scale publicly available data is at hand, as in the case of analyzing non-diversifiable market risk.

The transition towards a networked digital society has both increased the source and sophistication of risks and enabled the analyst to model them using sophisticated methods and approaches. Network analysis is a relatively new addition to the risk analytic toolbox but may prove to be an invaluable tool for modeling an even more complex world. Different approaches to understanding networks—purely visual, through network metrics, through comparison to idealized models, data-driven SNA and probabilistic BBNs—are still divergent. Researchers and practitioners in different field and traditions tend to prefer specific methods and tackle problems within their domain. It is likely that the future of risk management will draw upon the convergence of methods and paradigms to present an integrated network view of all the risks that face human, legal, or digital agents in their operations and interactions.

References

1. Backstrom, L., Boldi, P., Rosa, M., Ugander, J., Vigna, S.: Four degrees of separation. In: Proceedings of the 4th Annual ACM Web Science Conference, pp. 33–42 (2012)
2. Barabási, A.L.: Network Science. Cambridge University Press (2016)
3. Barabási, A.L., Albert, R.: Emergence of scaling in random networks. Science **286**(5439), 509–512 (1999)
4. Barefoot, K., Curtis, D., Jolliff, W., Nicholson, J.R., Omohundro, R.: Defining and measuring the digital economy, p. 15. US Department of Commerce Bureau of Economic Analysis, Washington, DC (2018)
5. Bindu, P.V., Thilagam, P.S.: Mining social networks for anomalies: methods and challenges. J. Netw. Comput. Appl. **68**, 213–229 (2016)
6. Bøttcher, S.G., Dethlefsen, C.: Deal: a package for learning Bayesian networks. Aalborg University, Department of Mathematical Sciences (2003)
7. Brugere, I., Gallagher, B., Berger-Wolf, T.Y.: Network structure inference, a survey: motivations, methods, and applications. ACM Comput. Surv. (CSUR) **51**(2), 1–39 (2018)

8. Can, U., Alatas, B.: A new direction in social network analysis: Online social network analysis problems and applications. Physica A: Statistical Mechanics and its Applications, **535**, 122372 (2019)
9. Castells, M.: The Rise of the Network Society, vol. 12. John wiley & sons (2011)
10. Chaney, T.: The network structure of international trade. Am. Econ. Rev. **104**(11), 3600–3634 (2014)
11. Cifuentes, R., Ferrucci, G., Shin, H.S.: Liquidity risk and contagion. J. Eur. Econ. Assoc. **3**(2–3), 556–566 (2005)
12. Cowell, R.G. (Ed.).: Probabilistic Networks and Expert Systems: Exact Computational Methods for Bayesian Networks. Springer Science & Business Media (2006)
13. Daly, R., Shen, Q.: Methods to accelerate the learning of Bayesian network structures. In: Proceedings of the 2007 UK Workshop on Computational Intelligence, Imperial College, London (2007)
14. De Andrade, R.L., Rêgo, L.C.: The use of nodes attributes in social network analysis with an application to an international trade network. Phys. A **491**, 249–270 (2018)
15. De Benedictis, L., Nenci, S., Santoni, G., Tajoli, L., Vicarelli, C.: Network analysis of world trade using the BACI-CEPII dataset. Glob. Econ. J. **14**(3–4), 287–343 (2014)
16. Freeman, L.: The Network Data Package. University of California Irvine, US (2009)
17. Gai, P., Kapadia, S.: Contagion in financial networks. Proc. Roy. Soc. A Math. Phys. Eng. Sci. **466**(2120), 2401–2423 (2010)
18. Gai, P., Kapadia, S.: Networks and systemic risk in the financial system. Oxf. Rev. Econ. Policy **35**(4), 586–613 (2019)
19. Gai, P., Haldane, A., Kapadia, S.: Complexity, concentration and contagion. J. Monet. Econ. **58**(5), 453–470 (2011)
20. Gregoriou, G.N., Lhabitant, F.S.: Madoff: a flock of red flags. J. Wealth Manage. **12**(1), 89–97 (2009)
21. Hautsch, N., Schaumburg, J., Schienle, M.: Forecasting systemic impact in financial networks. International Journal of Forecasting, **30**(3), 781-794 (2014)
22. Helms, R., Ignacio, R., Brinkkemper, S., Zonneveld, A.: Limitations of network analysis for studying efficiency and effectiveness of knowledge sharing. Electron. J. Knowl. Manag. **8**(1), 53 (2010)
23. Højsgaard, S.: Graphical Independence Networks with the gRain Package for R. J. Stat. Softw. **46**(10), 1–26 (2012)
24. Hurt, C.: Evil has a new name (and a new narrative): Bernard Madoff. Mich. St. L. Rev., 947 (2009)
25. Kaur, R., Singh, S.: A comparative analysis of structural graph metrics to identify anomalies in online social networks. Comput. Electr. Eng. **57**, 294–310 (2017)
26. Kirchner, B.: The Bernard Madoff Investment Scam. FT Press (2010)
27. Korb, K., Nicholson, A.E.: Bayesian Artificial Intelligence, 2nd edn. Chapman & Hall/CRC (2010)
28. Krugman, P.: The Madoff economy. New York Times **19** (2008)
29. Lauritzen, S.L.: Some modern applications of graphical models. Oxf. Stat. Sci. Ser. **27** (2003)
30. Liu, X., Li, D., Ma, M., Szymanski, B.K., Stanley, H.E., Gao, J.: Network resilience (2020). arXiv preprint arXiv:2007.14464
31. Margaritis, D.: Learning Bayesian network model structure from data. School of Computer Science, Carnegie-Mellon University, Pittsburgh, PA. Technical Report CMU-CS-03-153 (2003)
32. May, R.M., Arinaminpathy, N.: Systemic risk: the dynamics of model banking systems. J. R. Soc. Interface **7**(46), 823–838 (2010)
33. Merton, R.K.: The Matthew effect in science: the reward and communication systems of science are considered. Science **159**(3810), 56–63 (1968)
34 Milgram, S.: The small world problem. Psychology today, **2**(1), 60-67 (1967)
35. Newman, M.E.: Power laws, Pareto distributions and Zipf's law. Contemp. Phys. **46**(5), 323–351 (2005)

36. Newman, M.E.: Networks. Oxford University Press (2018)
37. Paul, G., Sreenivasan, S., Stanley, H.E.: Resilience of complex networks to random breakdown. Phys. Rev. E **72**(5), 056130 (2005)
38. Schneider, F., Feldmann, A., Krishnamurthy, B., Willinger, W.: Understanding online social network usage from a network perspective. In: Proceedings of the 9th ACM SIGCOMM Conference on Internet Measurement, pp. 35–48 (2009)
39. Scott, J.: Social network analysis: developments, advances, and prospects. Soc. Netw. Anal. Min. **1**(1), 21–26 (2011)
40 Scutari, M.: Learning Bayesian Networks with the bnlearn R Package. Journal of Statistical Software, **35**(i03), (2010)
41. Scutari, M., Graafland, C.E., Gutiérrez, J.M.: Who learns better bayesian network structures: Constraint-based, score-based or hybrid algorithms? In: International Conference on Probabilistic Graphical Models, pp. 416–427. PMLR (2018)
42. Smith, P., Hutchison, D., Sterbenz, J.P., Schöller, M., Fessi, A., Karaliopoulos, M., Lac, C., Plattner, B.: Network resilience: a systematic approach. IEEE Commun. Mag. **49**(7), 88–97 (2011)
43. Srivastava, A., Pillai, A., Gupta, D.J.: A walk through social network analysis: opportunities, limitations, and threats. In: Graph Theoretic Approaches for Analyzing Large-Scale Social Networks, pp. 1–17. IGI Global (2018)
44. Stattner, E., Vidot, N.: Social network analysis in epidemiology: current trends and perspectives. In: 2011 Fifth International Conference on Research Challenges in Information Science, pp. 1–11. IEEE (2011)
45. Strogatz, S.H.: Exploring complex networks. Nature **410**(6825), 268–276 (2001)
46. Tang, L., Liu, H.: Community detection and mining in social media. Synth. Lect. Data Min. Knowl. Discov. **2**(1), 1–137 (2010)
47. Tanizawa, T., Paul, G., Cohen, R., Havlin, S., Stanley, H.E.: Optimization of network robustness to waves of targeted and random attacks. Phys. Rev. E **71**(4), 047101 (2005)
48. Van Ark, B.: The productivity paradox of the new digital economy. Int. Prod. Monit. **31**, 3 (2016)
49. Wang, C.H., Cheng, H.Y., Deng, Y.T.: Using Bayesian belief network and time-series model to conduct prescriptive and predictive analytics for computer industries. Comput. Ind. Eng. **115**, 486–494 (2018)
50. Watanabe, C., Naveed, K., Tou, Y., Neittaanmäki, P.: Measuring GDP in the digital economy: increasing dependence on uncaptured GDP. Technol. Forecast. Soc. Chang. **137**, 226–240 (2018)
51. Watts, D.J.: A simple model of global cascades on random networks. Proc. Natl. Acad. Sci. **99**(9), 5766–5771 (2002)
52. West, G.: Scale: the universal laws of life and death in organisms. Cities Co. (Weidenfeld & Nicolson, 2017)

Chapter 5
Analyzing Rare Risks

5.1 Introduction

Managing rare risks is the Holy Grail of risk management. Those rare risks are extremely unlikely events with potentially catastrophic impact. This makes their prediction devilishly difficult, yet supremely important as they often mean the difference between life and death of individuals, companies, and even whole systems. Makridakis[1] aptly described the difference between normal mundane risks and those rare calamities. The former, he surmised, are subway risks. As a given individual walks every morning to the subway, she may be beset by a number of uncertain events, ranging from excessive traffic to subway delays. All these events have a noticeable but marginal impact on travel time, tend to cluster towards a central point and are relatively randomly but symmetrically dispersed around it. Until one day a heavy object falls on the individual's head, and this one risk changes everything with potentially huge impact on travel time but also health and even life. This is a coconut risk.

Nassim Taleb[2] used yet another metaphor, inspired by European settlers discovering new lands and species during their early explorations. Seamen who disembarked in Australia and were used to only white swans were shocked to see a black swan inhabiting those lands. A black swan, according to Taleb, is a rare, and potentially catastrophic event, that no one could have foreseen due to lack of imagination and available data. Such a black swan may be the rise of the Internet or a severe economic recession such as the Great Depression. Some events are foreseeable but unexpected by most—those risks Taleb termed gray swans. And finally, the white swans are the known and tractable risks that can be managed in a straightforward way. While these are easiest to tackle, it is precisely others—the elusive swans of darker color—that are crucial at a systemic level. This chapter is all about understanding those and creating management strategies for a more sustainable world.

---

[1] Makridakis et al. [29].

[2] Taleb [46].

A. Gerunov, *Risk Analysis for the Digital Age*, Studies in Systems, Decision and Control 219, https://doi.org/10.1007/978-3-031-18100-9_5

It is worth noting that in order to manage such rare risks, the analyst must be able to predict them with some degree of confidence. Wright and Goodwin propose a key distinction in the meaning of the term predictability.[3] In its first sense predictability means creating a well-calibrated probability distribution of the variable under interest. An example may illustrate this concept. If a weather forecaster produces 100 forecasts with a 20% probability of rain, one would expect 20 rainy days if the probability distribution is well-calibrated. The second sense of predictability builds upon our human need to decrease uncertainty and the business imperative for clarity. It embodies the idea that the generated forecast would be close enough to the actual realization. Thus, a phenomenon, like sales revenue, is deemed more predictable if a generated forecast has a standard deviation of 1 Million than if it had a standard deviation of 3 Million of some currency. Decision theory clearly shows that predictability of the first type is necessary and sufficient for rational decisions, and practice underline its use in devising concrete mitigation strategies. When it comes to unlikely events, this is also the best one can hope for, and thus for good theoretical and practical reasons we focus on it.

The key issue with rare events is exactly that—this rarity means we have a very limited amount of knowledge and data (sometimes zero) that can be used to forecast their probability of occurrence and impact. The general dearth of information is an issue across all forecasting approaches—all the way from expert judgment with no precedent to rely upon to sophisticated statistical models that have no data to be fed with. Yet, the potentially huge impact of black swans engenders the clear imperative of managing them, and quite a few approaches to understanding rare events have proliferated.

Still, obtaining data for events that (almost) never happen is a daunting challenge and methods for analysis are plagued by a number of challenges[4] as follows (see Fig. 5.1):

- **Sparsity of reference class**—there are very limited or sometimes even no observations of the relevant event, thus making inference about probability or impact very challenging.
- **Inappropriate or incomplete reference class**—collecting data on analogous events may be misplaced as those analogies could be rather imperfect or incomplete.
- **Inappropriate statistical model**—sometimes an erroneous model, statistical distribution, or underlying assumptions may pollute the analysis. With sufficient information, the risk manager is able to overcome those pitfalls by letting the data speak. This is hardly the case with rare risks,
- **Misplaced causality**—again a data problem, seeing causal links where there are none and lacking data to check this assumption may skew and bias the results.
- **Frame blindness**—refers to the fact that any analytic exercise operates within a given paradigm, under specific assumptions and constraints, and within the

[3] Wright and Goodwin [54].

[4] Goodwin and Wright [16].

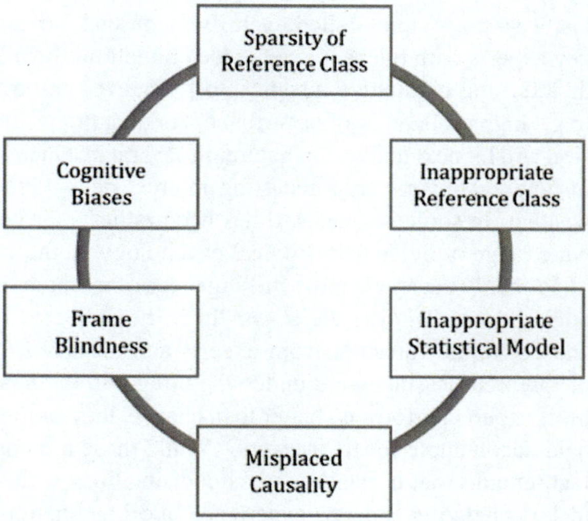

Fig. 5.1 Key sources of error for forecasting rare events, based on Goodwin & Wright, [16]

context of accumulated knowledge and conventions. This may make the analyst blind to alternative information or explanation, thus compromising the analytic process.
- **Cognitive biases**—those refer to the many biases and imperfections that humans exhibit when making forecasts and decisions in situations of uncertainty.

We will briefly review all analytic approaches, starting from expert judgment (group or individual), statistical methods, the more novel approach of prediction markets, and using scenarios and Monte Carlo simulations. The chapter then presents a novel approach combining the strengths of expert judgment, statistical modeling, and Monte Carlo simulation to gain a better understanding of rare risks in the digital economy. The aim is to outline the limits of predictability of rare risks and yet create a meaningful framework that allows for the evaluation of their potential scope. Based on that, contingency management plans can be devised within the constraints of organizational resources and pre-defined risk appetite. The take home message is that precise forecasting of extremely rare events is impossible but a structured approach to managing unlikely eventualities is crucial in modern business and society.

5.2 Expert Judgment

Using expert judgment to aid decision-making in situations of uncertainty has long preceded the formal discipline of risk management. Dating back to tribe elders, deferring to subject matter expertise has been part and parcel of individual and organizational decision-making. In fact, the first risk management consultants known in history—the Asipu—relied exclusively on that.[5] It essentially includes identifying key experts with relevant experience and letting them estimate variables of interest, thus turning qualitative insight into formalized information that can be qualitative (e.g. "high probability of occurrence") or quantitative (e.g. "90% certainty that recession will hit next quarter) in nature. In several instances this has been quite successful in fields as diverse as forecasting interest rates, horse racing, card games, or the weather. In some instances, it has been rather poor with research again covering a wide range of fields from clinical psychology all the way to maize judgments.[6]

Probably the single most striking piece of research in the field is Gordon Tetlock's book *Expert Political Judgment.* In it, he collected no less than 82,361 political and economic forecasts from experts and observed their realization over decades of research. Results were underwhelming—in the fields of politics and economics most experts perform no better than chance, they (almost) never learn, and are rarely held accountable for their errors.[7] While those are compelling insights indeed, the fact remains that in poorly understood situations with no data, often the only source of knowledge is experts' experience and imagination. Thus, expert judgment has remained a central approach to obtaining probability and impact estimates for rare events. Errors in human decision-making are studied assiduously,[8] and researchers have proposed ways to structure the decision situation or the outcomes in such a way as to minimize biases and increase accuracy. The quest for this is far from complete.[9]

5.2.1 Individual Judgment

The classic way to glean information from experts is to rely on their individual unadulterated judgment. They are thus asked to identify potential risks and evaluate their probability or impact on a qualitative or a quantitative scale. The resulting evaluations can then be summarized in a risk matrix that enables the analyst to prioritize risks.[10] Those forecasts, as in Tetlock, are often found to be wanting due to a number of compelling reasons[11]:

- **Sparsity of reference class**—there may be insufficient data on the type of event under investigation, thus making it impossible to estimate its characteristics. This holds particularly true for rare risky events. In general, the broader the definition of the reference class, the more instances one is likely to observe. For example, total accidents in nuclear power plants is a much broader reference class than nuclear

[5] For the story of the Asipu, please see Chap. 1.

[6] See original references in Goodwin and Wright [16], p. 360.

[7] Tetlock [47].

[8] Kahneman [24].

[9] See e.g. Hemming et al. [21], van Dijk and Franses [49], Aromi [3].

[10] See Ni et al. [34], Bao et al. [6].

[11] Goodwin and Wright [16].

incidents, caused by a tsunami following a catastrophic earthquake. Even though the latter is exactly what happened in the Fukushima Daiichi Nuclear Power Plant in March 2011, the former is what is much more amenable to prediction. It is thus much more feasible to predict rare risks in terms of larger categories instead of very specific events.

- **Inappropriate or incomplete reference class**—when modeling extremely rare events, the observations at hand are merely samples of the population of possible realizations. They thus may not include relevant rare events (i.e. it has not yet happened, but it is nonetheless possible) or be biased towards common and predictable events. This can lead to poor calibration of forecasts, leading the analyst astray. A possible way to avoid this is by basing the management of rare risks not only on recent actual scenarios but also on remote and hypothetical ones. Taking issue with the Fukushima Daiichi accident, one must note that earthquakes and tsunamis of comparable proportions did happen at least twice in recorded history in the ninth and seventeenth century.[12] On the other hand, risk analysis only took account of the recent past. When it comes to the digital economy, and the information assets it creates, it is a viable alternative to also build hypothetical scenarios based on process drivers (e.g. create a risk model from first principles), irrespective of whether the actual event modeled is realized or not. Modeling the risk and impact of large-scale introduction of automated decision-making (i.e. Artificial Intelligence) is an example of such a case.
- **Misplaced causality**—oftentimes human experts have a pre-determined conception of the causal chains in a given complex system. However, even a broad scientific and expert consensus in the field does not guarantee the existence of those clear deterministic causal links. More insidiously, existing data may give initial support to a given theory by displaying high correlations between variables and robust linear dependence between them, thus leading to a wide adoption by the mainstream. This alone does not prove causality as well since one may be dealing with spurious correlation and subjectivity of interpretation. What is more, human experts may have a pet theory they espouse, and implicitly bring it into the evaluation and forecasting tasks. In view of the digital economy that is a relatively recent and still under-researched phenomenon, this source of error may be of importance.
- **Cognitive biases**—the final group of reasons for errors stems from the very way the human expert's brain functions. Kahneman and Tversky's pioneering research has found a set of consistent heuristics and biases that people use to estimate probability, and these lead to systematic errors.[13] For example, the representativeness heuristic leads expert to assign much higher probability to events that are recent, unusual, or particularly vivid in the brain of the estimator. In this vein, the Fukushima earthquake and tsunami were probably assigned de-facto zero probability even though those events could plausibly happen (and indeed did). The representativeness heuristic leads human astray by letting them believe that

[12] Pate-Cornell [39].

[13] Kahneman [24].

a certain event is representative of a given class and ignoring basic unconditional probabilities. Finally, the anchoring heuristic means that estimates are anchored to some number (even an arbitrary one) and fail to make sufficient adjustment. Estimating quantities related to the digital economy is likely to be subject to all three, and thus care must be taken to design appropriate elicitation techniques, and possibly train the experts before elicitation.

Irrespective of the precise source of error, in the final analysis, one finds that even among experts working in the same close-knit field, there may be significant disagreement, thus introducing large uncertainty in the forecast.[14]

5.2.2 Judgment Decomposition

In an effort to address the shortcomings of individual judgment, experts have introduced different ways to structure the judgment situation such that known biases are minimized. One of those is the structured judgment decomposition which aims to deconstruct the forecasting task into smaller and more manageable sub-tasks. Since the problem becomes somewhat more tractable, it is also likely that the quality of estimates of probabilities and effects should tend to increase.[15] Decomposition thus crucially depends on eliminating the oversimplifying heuristics in human judgment in order to reduce the biases stemming from them. In this respect, using tools such as event (decision) trees or fault trees may be of particular use.

The tree allows the estimation of a precursor events and a clear causal link between initial events and the risky outcomes. This in turn makes the assumptions explicit and allows the resulting probability of the rare event to be estimated by aggregating intermediate probabilities. Those may be obtained either as a result of expert judgment or through available statistical data. Furthermore, separating probability estimation from outcome projection may decrease the effect of wishful thinking or optimism bias, and be used as an antidote against group thinking. Furthermore, this forces participants to be explicit about their ideas and assumptions and to consider cases that they would not have otherwise but now have to as these are on the causal chain, as well.[16]

Some issues with decomposed expert judgment still remain:

- **Sparsity of reference class**—this source of problems is better tackled than in the case of individual judgment as the availability heuristic is addressed by enabling participants to cover all possibilities. A formal mathematical check is possible to ensure the consistency of estimates as the analyst aggregates from precursor events to the final rare event. This process also enables a sanity check whereby probability estimates elicited by experts are cross-checked against actual empirical

[14] Pasman and Rogers [38].

[15] Chang et al. [11].

[16] Goodwin and Wright [16].

data that may very well available for precursors as they are events with much higher frequency. As insufficient observation of the rare class of events will likely be a consistent feature of modeling emerging phenomena in the digital economy, the decreased scope of error here is a particularly welcome development.

- **Misplaced Causality**—a structured causal chain is as helpful as its reflection on actual empirical reality is truthful. Experts tend to have a shared belief of the structure of causality in their domain and vigorously defend it against all opposing views, and even some facts. Leveraging a misplaced causality chain tends to worsen forecasting and not improve it. Furthermore, probability aggregation from precursors to target events become unreliable if wrong precursors are chosen, or if they are placed in the wrong order. This rings particularly true for problems in the realm of the digital economy.
- **Frame blindness**—the issue of experts looking at a given forecasting tasks from their own preferred paradigm (frame) is arguably even exacerbated when the risk analyst goes down the road of judgment decomposition. In fact, the very decomposition itself is predicated on the experts' frame which enable the elicitation of causality links. This causes marked concerns in complex or under-researched fields such as the economic system and the intersection of business and technology.

Thus, while judgement decomposition addresses some of the specific concerns of counting on unreliable human estimates, it is hardly a comprehensive solution eliminating all potential sources of error. Its application merely enables the analyst to reduce error below the levels obtained in unadulterated spontaneous judgment. Whether this decrease in uncertainty reaches acceptable levels requires an answer for each separate decision situation.

5.2.3 Structured Analogies

A well-known bias observed in human decision-makers is the availability heuristic. It refers to the fact that when asked to provide an estimate, experts think of similar occurrences and make an inference based on those events. However, they are more likely to recall an event of higher salience that literally jumps to their minds. In short, estimates are driven by the most easily available comparisons.[17] Knowing that such comparisons are unavoidable when eliciting evaluations from experts, the deft analyst may choose to provide reference cases in advance so that the experts are exposed to relevant and representative comparisons and not merely left to the chance occurrence that relevant analogies stand on top of their head. By doing this the risk manager improves the accuracy of elicited responses by diminishing the effect of the availability heuristic.

[17] Redelmeier and Ng [41].

In exactly such a setup Lee et al.[18] looked into ways of improving estimates on the effects of future sales promotions. Experts were provided with a database that contains information on previous promotional campaigns effectively making this the available analogy for estimates of the future campaign. Furthermore, experts were given an idea of how similar past data is to current promotion, and the observed effects on demand. Unsurprisingly, this led to a significant improvement in the generated forecasts—a result that carries over to other research on structured analogies.[19]

Using structured analogies addresses the issue of the sparsity of the reference class. By selecting available similar cases the analyst achieves an effective widening of the set of events under study and thus improves the estimates of their characteristics—both in terms of impact and of probability. Bias reduction is another major strength of this approach as it reduces biases of recall and aptly selected analogies may even improve representativeness. On the other hand, some problematic issues persist, as follows:

- **Inappropriate or incomplete reference class**—when choosing an analogy, the analyst must have a high degree of confidence that it is, in fact, an apt one. If an imperfect analogy is selected this may confuse the experts leading them astray. Framing them with erroneous cases or events is likely to deteriorate the quality of both impact and probability estimates as experts evaluate the parameters of a very different phenomenon. This risk is particularly important in problems of the digital economy where technology and business context specificities may lead to unique cases with no clear counterparts.
- **Misplaced causality**—another root cause of using inappropriate analogies would be reliance on false causal assumptions. Misunderstanding of driving forces and their effects may lead to seeing similarities when there are none, and priming experts to see and extrapolate them in their evaluations. Thus, care must be taken to ensure that the choice of relevant reference cases is based on solid subject matter expertise, proven causal links, and sufficient representativeness.
- **Frame blindness**—a last issue that is not addressed by the structured analogy method is the frame blindness. If anything, the method exacerbates the tendency to neatly fit disparate phenomena into well-known and understood patterns.

In short, the method of structured analogies is only as good as the analogy itself. If it is an apt one, the expert will be primed to it, and anchored to its qualitative characteristics and quantitative implications. Thus, the forecast for a potentially rare and catastrophic event will improve. If the analogy selection is a failure, the expert is bound to be no less primed and anchored, but the estimates produced may end as a disaster.

---

[18] Lee et al. [26].

[19] Green and Armstrong [18], Nikopoulos et al. [35].

5.2.4 Group Judgment Methods

Individual decisions are fickle and unreliable, but some authors have argued that one may find surprising clarity and accuracy in groups of individuals. This phenomenon, known as the wisdom of the crowds, has captured the imagination of many researchers, practitioners, and pundits, hailing it as a crucial tool for all sorts of group decision-making.[20] The intuition behind the power of crowds is enticingly simple. When a given individual makes an estimate, it is influenced by the decision architecture and skewed in a certain direction by the heuristics and biases this individual is exposed. However, different individuals tend to suffer from random disturbances in their estimates and thus when aggregating individual forecasts those random errors cancel each other out, thus leaving the analyst with an accurate number. This has led to the proliferation of group judgment methods in all walks of life but one needs to take heed that only some of them are suitable for risk elicitation.[21]

The simple group of interacting individuals (focus group) is particularly suspect. For the wisdom of the crowd to work, individual estimates (and thus experts) must be independent from each other while groups tend to form consensus opinions. This phenomenon, the "groupthink", is what compromises accuracy as the errors stop being random and instead are biased in one and the same direction. It is hardly surprising that simple group estimates significantly underperform alternative viable methods.[22] One of the major contenders for a structured group judgment method in forecasting and risk management is the Delphi method.

It involves collating a group of experts who are asked to give their estimates in a first round of forecasting. These anonymized individual estimates are then shared with other members of the set. In addition to exchanging numeric projections, experts may also exchange reasons for their expectations. In subsequent rounds, experts are allowed to change their opinion and update their forecast, possibly sharing their motivation to do so. The Delphi panel continues iterating as the participants reach a consensus and the group feels comfortable with both probability and impact estimates. The Delphi procedure has proved to be more accurate as compared to alternatives and tends to outperform simple statistical groups whereby individual judgments are mechanically aggregated to reach the group estimate.[23]

The reasons behind the effectiveness of the Delphi method are still subject to research but it is most often argued that the ability to exchange anonymous opinions is key to improving the forecast of the rare and risky event. This help lessen the effect of inappropriate framing of the issue and addresses a few key problems when generating forecasts. Most notably, it improves the recall of similar cases, lessening the importance of the availability heuristic and ensuring the proper reference class is being used across the panel. Moreover, the exchange of expert opinions can lead

---

[20] Surowiecki [45].

[21] Hertwig [22].

[22] Armstrong [2].

[23] Rowe and Wright [44], Phadnis [40].

to more critical scrutiny of the frame of reference and expose the use of pet theories. This naturally helps ameliorate the effects of the biases connected to frame blindness.

Despite its many advantage, there are some outstanding issues when using the Delphi method, that the analyst must be aware of. Most notably, this is the **misplaced causality** problem. A panel of experts may trade arguments based on specific theory (e.g. the state-of-the-art in the their respective field) which however assigns erroneous causal links to the events that are being forecasted and their drivers. This strain of misplaced causality may bias and skew the estimates, thus worsening the information obtained from the Delphi method. At any rate, one needs to keep in mind that any information elicitation method will have its problems and downfalls but those need to be balanced against its benefits is a rational way. Thus, the most important task standing before the analyst is to make a judicious choice of elicitation should judgement calls be unsurmountable, and to correct for known problems in the method used.

5.2.5 Uncertainty and Calibration

Judgment by individual human experts is crucially influenced by the elicitation architecture and the particularities of the experts themselves. This well-known fact is hardly an insight for the seasoned analyst but it puts with particular poignancy the question of how to approach and rectify the errors in the elicited estimates, and how to account for the uncertainty inherent in those projections. Tackling this begins in the very structure of obtained response—and there are five principal approaches to eliciting expert opinion, as follows[24]:

- **Point values**—the expert is asked to provide a single point estimate of the quantity of interest, e.g. the probability of event A happening. While these are easy and intuitive to collect, they are subject to some significant disadvantages. Most notably, the analyst is unable to reconstruct the probability distribution of the variable under study since a single point on it gives little information. A point estimate also hides the uncertainty surrounding the projection and this may have significant quantitative impact. Finally, individual point estimates by experts sometimes tend to be operated on as if they were physical quantities, and they are most certainly not. Estimates are merely the best guess for a realization of a random variable.
- **Paired comparisons**—in this case the experts are asked to compare and rank alternatives in order to feed ranks into a decision-making algorithm. This is intuitively simple and harks back to the paired comparison of choices so prominent in utility theory. This is tractable when the expert is presented with a limited number of alternative but as they grow, the number of comparisons needed also starts growing to overwhelming proportions. Paired ranks also do not allow one to get

[24] Goossens and Cooke [17].

an idea of the uncertainty surrounding the estimate, or to construct the relevant statistical distribution that may be needed for further analysis.

- **Discrete event probabilities**—this mode of elicitation asks experts to provide a determination of an event occurring in a binary form—i.e. if the event is going to occur or not. An example from finance would be to ask experts whether a given stock market will go up over the next year or not. Such discrete probabilities yield themselves to formal mathematical modeling and may be subject to the tools of probability theory. In fact, some random variables, such as the coin toss, are binary in nature and so are some events in real life. This elicitation is thus particularly suitable for them.
- **Distributions of continuous uncertain quantities**—this elicitation results in a large quantity of information to the risk analyst that may be processed and analyzed. Experts provide information so that the complete distribution of a continuous variable of interest may be estimated or reconstructed, essentially giving ideas of their subjective distribution. The greatest strength in this approach lies in the ability to quantify the estimate uncertainty and be able to run simulations and sensitivity analyses on the data. Subject matter experts may not always have extensive knowledge in statistics which is why it is common to ask them to provide only a few parameters that can be used to reconstruct the distribution such as the mean and standard deviation for Gaussian phenomena, or quantiles that can be used to fit it. In practice, it may be enough to elicit just the 5% and the 95% quantiles.[25]
- **Conditional probabilities**—a final and most sophisticated way to elicit expert knowledge is to contextualize the variable that needs to be evaluated. It may very well be the case that this variable depends on a number of other factors, and experts need to be made aware of the known realizations of those factors. This way they do not provide the unconditional probabilities for an event happening but rather the conditional ones based on the occurrence of other related events. From a psychological standpoint, it is likely that experts will try to conditionalize the estimate themselves but may lack the information to do so, and thus errors will be introduced. The importance of leveraging conditional probabilities can hardly be understated and their availability allows a much richer analysis by taking recourse to Bayesian methods and ideas.

From the standpoint of research and practice, the analysis of risks only benefits from a larger trove of available information and thus the elicitation of information that leads to the construction of probability distributions is much preferred.

Whatever the way of elicitation, the resulting estimates must be treated with caution. While a few studies show that estimates could be rather accurate, the overwhelming majority of research has shown that experts are overconfident and provide estimates with a significant degree of error.[26] This issue is known as expert calibration and as particularly pertinent when dealing with uncertainty estimates. A quick

---

[25] Hubbard and Drummond [23].

[26] Lin and Bier [27], McBride et al. [33].

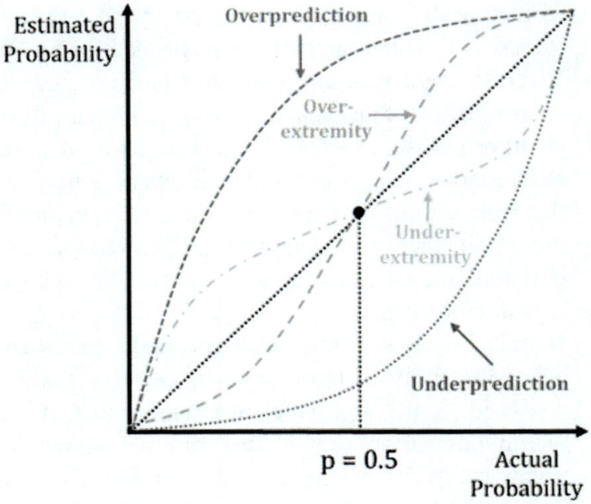

Fig. 5.2 Types of Mis-calibrated Estimates, source: Koehler et al., [25]

check for overconfidence is asking experts to provide a confidence interval such as a 90% in which the true value must lie. The analyst then checks whether experts are right in 90% of the cases, which is most often not observed.[27] While there are many more sophisticated ways to evaluate calibration, this is a standing issue that plagues expert opinion elicitation.

In principle there are four major types of behaviors when forecasting events that are difficult to estimate in complex environments (such as rare risks). Those are as follows and appear in different forecasting situations with different types of experts and non-experts[28]:

- *Overprediction bias*—the expert systematically generates estimates that are significantly above actual realization;
- *Over-extremity bias*—in this case the prediction is very conservative below probabilities of 50%, and overly optimistic when probability is above 50%;
- *Under-extremity*—this is the mirror image of the over-extremity bias, in which experts overestimate low-probability event of below 50%, and underestimate high-probability events;
- *Underprediction bias*—experts with this bias systematically provide estimates that are below the actual realizations, showing extreme conservatism (Fig. 5.2).

While underprediction and overprediction bias are intuitive to understand, over- and under-extremity can be surprising at first. Their psychological root is the inflection point in how people perceive probability and was amply investigated by behavioral psychology. The results, in short, reveal that the utility function of individuals

[27] Lin and Bier [27], Velasquez et al. [51].

[28] Koehler et al. [25].

is skewed around certain points which reflects the miscalculation of objective probability.[29] Thus, essentially what one observes in complex and unpredictable environments is the phenomenon of experts that provide wrong estimates with great confidence. Some studies even show that some students can beat experts in prediction tasks of significant complexity such as predicting the outcomes of ecological research.[30]

The single best solution for the analyst is to calibrate experts and the scope of expert calibration may be very different across domains and study iterations, and also the amount of noise may in some instances drown the useful signals elicited.[31] What is more, part of the calibration procedures may enable the analyst to pinpoint experts with better forecasting accuracy and accordingly give their prediction a greater weight in the final aggregation. This can be done by using seed questions, whereby experts are asked to estimate variable unknown to the expert but known to the analyst and the percent correct responses are used to rank the forecasting skill of the experts.[32]

The actual calibration may proceed in many fashions—the analyst may choose to calibrate experts themselves, or to adjust results afterwards. While the former is key, nothing precludes one from applying both. Individual expert calibration includes assessing both their competence and their ability to provide accurate estimates.[33] The latter is judged by asking seed questions such as estimating the fraction of days that the stock market has gone up over the past five years and the interval of values that contain the actual number with 90% certainty. Those known quantities from these seed questions are then used to estimate calibration and even train the experts further until they are well-calibrated and able to provide accurate estimates.[34] The metaphor usually used for this process is that similar to how engineers calibrate their instruments or musicians fine-tune theirs, so the risk analyst must enable the calibration of the estimates used.

At this point, one may pause and ask the question of why experts are used at all since plenty of research has cast doubt of their abilities to generate useful estimates, with some saying they are not much better than chance.[35] On top of this, the analyst may be left wondering as to the direction of the bias with the only thing being for sure is that the experts themselves are truly confident of what they said. The key issue with using expertise is that in the case of rare risks it is often the case that sufficient statistical data is not available, and thus the process of expert elicitation is merely a process of turning well-founded imagination into numeric estimates that can feed the analytic process. Experts are thus unsurmountable but need to be selected carefully, vetted vigorously, trained thoroughly, and weighted accordingly lest their estimates fool not only themselves but the risk manager, too.

---

[29] O'Donoghue and Somerville [36].

[30] McBride et al. [33].

[31] Walker et al. [52], Lin and Bier [27].

[32] See e.g. Goossens and Cooke [17], Grigore et al. [19].

[33] Unal et al. [48].

[34] Hubbard and Drummond [23].

[35] Tetlock [47].

5.3 Statistical Methods

The use of historical data to feed a more or less sophisticated statistical algorithm is part and parcel of economic forecasting. Forecasting problems abound in all fields of economic inquiry, ranging from macroeconomic issues, through financial markets dynamics, and all the way into microeconomic problems such as consumer choice and demand planning, to name a few. Despite the growth of data and the advancement of methods researchers and analysts find in many instances that forecasts remain wanting.[36] The occasional laments notwithstanding, estimating the future is now a well-researched and somewhat commoditized problem that yields itself to a set of well-established approaches, methodologies, algorithms and statistical estimation procedures, as is well exemplified in the M series of forecasting competitions.[37] Even simple methods are now able to provide satisfactory estimates, and ensembles of methods do an even better job, but they all generate a certain unavoidable level of errors. Still, the forecasts are good enough in the overwhelming majority of cases with one notable exception—when it comes to forecasting truly rare events. This specific case calls for a deft combination of data, insights from probability theory, and expert judgment.

5.3.1 Pure Statistical Methods

Pure statistical methods tend to use a large amount of data and process the information contained therein so as to produce a best estimate of future realization of a target variable of interest. While many classifications of statistical methods and models do exist, one needs to make the distinction between the following[38]: time series methods, structural models, and ensemble approaches. Each of them has its own specific strengths and weaknesses but all derive their quality from the quality and quantity of input data.

Time Series Forecasting Models—in these the variable of interest is forecasted using the information contained in its previous realizations (lags). The assumption here is that previous values of the target variable hold important characteristics of its dynamics such as its trend and cyclicality, and these can be exploited to generate a forecast. Typically, the analyst resorts to modeling lagged values (autoregressive terms), averaged model errors (moving averages) and makes corrections to ensure that the time series is stationary. Those considerations give rise to family of very popular time series models the Autoregressive Integrated Moving Average (ARIMA) models that are in common use. A common form of the ARMA model for a dependent variable y_t and previous errors $\varepsilon_{t\text{-}i}$ is as follows (α_i, β_i and coefficients):

---

[36] Orrel and McSharry [37].

[37] Makridakis et al. [31].

[38] Makridakis et al. [30], Hastie et al. [20].

$$y_t = \beta_0 + \sum_{i=1}^{p} \beta_i y_{t-i} + \sum_{i=1}^{q} \alpha_i \varepsilon_{t-i} + \epsilon_t \tag{5.1}$$

If the time series is not stationarity it may be made so by differencing the dependent variable. Running the same model as in Eq. (5.1) on the series of once-differenced y_t makes the model integrated of order 1. Should the time series need n differences for stationarity, the analyst is said to run an ARIMA integrated of order n. Popular other time series methods include different extrapolations and exponential smoothing approaches. Even simple time series model are found to generate relatively accurate projections which partly explains their popularity. However, it is often the case that rare events feature only sparsely in the time series (or even not at all) and thus the models are unable to capture and accurately forecast them. Time series models are thus best-suited for short-term forecasting in stable environments.

Structural Forecasting Models—in those a set of independent predictors (explanatory variables) is used to forecast a dependent target variable. This type of models aims to exploit the structural relations between different variables, and to quantify their interdependence and effect. To this end, a causal relationship is often pre-supposed, which enables the modeler to estimate the effect of a change in one independent variable on the target variable. As all relevant changes are aggregated, one obtains the net effects on the target and so—its value. The simplest among structural models is the linear regression which relates the effects of a set of k independent variables x_i on a dependent variable y and quantifies them through the magnitude of the β_i coefficients:

$$y = \beta_0 + \sum_{i=1}^{k} \beta_i x_i + \varepsilon \tag{5.2}$$

Basic models such as the linear and logistic regression are well-known in economics and business but more complex relationships call for more sophisticated ones such as more complex members of the Generalized Linear Model (GLM) family. Additionally, novel methods from other quantitative fields such as the neural network, decision trees, or support vector machines are also increasingly gaining prominence. Structural models have a number of important drawbacks. Most importantly, they assume invariance of the relationship between variable which easily breaks down especially in the infrequent and calamitous circumstances that rare risks pose. On the practical side, in order to predict future realizations of the dependent variable, the analyst needs the future realizations of the independent ones, thus adding an additional layer of forecasting complexity. All this makes structural models very useful for understanding process drivers and system links but challenging to use in forecasting rare events when those links may behave differently.

Combined and Ensemble Models—looking to take the best from both worlds, analysts have amply used the opportunity to leverage models that include both time series and structural components. By doing so, they inherit the strengths of both approaches, but also their drawbacks. For example, more sophisticated models such as the Vector Autoregression (VAR) may take advantage of both temporal structure of the data, and causal links between variables studied in a simultaneous regression framework. The equations in (5.3) define a VAR model where the variables x_t, y_t and z_t are studied, whereby each of them is dependent both on its previous realizations and on the previous realizations of the other variables with strength of effects ϕ_{ij} (with c_i being a constant term):

$$x_t = c_1 + \phi_{11}x_{t-1} + \phi_{12}y_{t-1} + \phi_{13}z_{t-1} + \varepsilon_1$$

$$y_t = c_2 + \phi_{21}x_{t-1} + \phi_{22}y_{t-1} + \phi_{23}z_{t-1} + \varepsilon_2$$

$$z_t = c_3 + \phi_{11}x_{t-1} + \phi_{12}y_{t-1} + \phi_{13}z_{t-1} + \varepsilon_3 \tag{5.3}$$

There are many variations on this idea—from simplifying it to use a single equation to making it more complex by including more lags and correcting for cointegration as in the Vector Error Correction (VEC) family of models. At any rate, the idea behind combining time series and structural causal relations between variables is to improve forecast accuracy, especially in challenging tasks. Another approach aimed at this is to combine not different ideas within a single model but rather the results of different models in an aggregated ensemble model. A classic example of this is the random forest model that is the combination of numerous individual decision trees. The resulting output is the majority vote or other similar aggregation of individual tree outcomes.

Finally, the analyst may decide to combine the numeric results of different forecasting models by applying some aggregation procedure such as using a (weighted) average. Significant research has shown that such aggregations tend to outperform individual forecasts,[39] and sometimes even outperform the single best forecast. In fact, adding additional forecasts is so effective that the estimate may be improved merely by adding a white noise term. The intuition behind the power of aggregation is quite straightforward: different types of algorithms capture and exploit different facets of the data modeled. By combining them, the analyst generates a more comprehensive view of the generating process that includes all its peculiarities and is thus able to produce superior projections for future realizations. This can easily be illustrated within a straightforward mathematical framework leveraging two forecasts.[40] If there are two respective forecasts F_1 and F_2 for a given variable of interest y, then the combined forecast F_c can be expressed as follows (ω denotes respective weights):

$$F_c = \omega F_1 + (1 - \omega)F_2 \tag{5.4}$$

Denoting the forecast error as $e_i \equiv y - F_i$ and assuming that the forecasts are unbiased, it follows that:

---

[39] Makridakis and Hibon [28], Makridakis et al. [31].

[40] Van Dijk and Franses [48].

$$E[e_i] = E[y - F_i] = 0 \tag{5.5}$$

The combined forecast error e_c for Eq. (5.4) can be expressed as follows:

$$e_c = \omega e_1 + (1 - \omega)e_2 \tag{5.6}$$

The variance σ_c^2 of e_c is thus given by the following equation with σ_{ij} denoting covariance:

$$\sigma_c^2(\omega) = \omega^2\sigma_1^2 + (1 - \omega)^2\sigma_2^2 + 2\omega(1 - \omega)\sigma_{12} \tag{5.7}$$

Since individual forecasts are unbiased, so is the combined one Fc. It follows from this that the mean squared prediction error (MSPE) is equal to $\sigma_c^2(\omega)$. Minimizing error is thus equal to minimizing this variance term and can be achieved by setting an optimal weight ω. This optimal weight is thus found to be:

$$\omega^* = \frac{\sigma_2^2 - \sigma_{12}}{\sigma_1^2 + \sigma_2^2 - 2\sigma_{12}} \tag{5.8}$$

With the optimal weight in Eq. (5.8) one can also estimate the variance of the combined forecast by substituting it into Eq. (5.7). It is found to be the following:

$$\sigma_c^2(\omega^*) = \frac{\sigma_1^2\sigma_2^2 - \sigma_{12}^2}{\sigma_1^2 + \sigma_2^2 - 2\sigma_{12}} \tag{5.9}$$

Equation (5.9) gives an expression for the error of forecast when combining different estimates, and it can be easily shown that the resulting error (variance) $\sigma_c^2(\omega^*)$ is always smaller than σ_1^2 and σ_2^2 except when either of those is equal to their covariance. In the latter case the weight is either zero or one, and the variance of the combined forecast is equal to the individual variance. Thus, if the two forecasts are exactly the same, there is no benefit in combining. However, if there is even a small difference between them, combination yields a decrease in the forecast error.

Traditionally hailing from statistical theory, forecasting models tend to assume a relatively large number of data points on which to operate. Novel approaches stemming from the fields of computer science and machine learning are created for the world of exponentially expanding information (big data) and thus thrive on particularly large sample sizes. However, the realizations of rare risks are, by definition, rare, and thus limited data is available for reconstructing statistical distributions and using historical data for reliable inference. Thus, forecasting rare events using pure statistical methods face significant challenges in the following areas:

- **Sparsity of reference class**—few observations of the reference class essentially limit the sample size upon which the statistical procedure is performed. While hardcore statisticians ask for the stringent requirement of having large samples, rules of thumb such as having at least 30 observations seem to dominate practice.

However, this may still be an unattainably ambitious goal for rare events, and observed realization could be much less. In those cases neither the statistical distribution can be reliably built, nor a forecasting model can yield reasonable estimates. In fact, pure statistical methods may function worse than expert opinion in this case as they contain less information: experts know both the data and their opinion, while the method only uses data.

- **Inappropriate reference case**—due to the limited number of realizations, the analyst may choose to use data on analogous variables as a proxy for the rare risk. This can be onerous since the new reference class be inappropriate and thus all inferences based on it can be hugely erroneous. Such a course of action is particularly attractive in dealing with issues of the digital economy as some large-scale catastrophic events (e.g. loss of life due to automated decision-making) never materialize and thus data on seemingly comparable events can be leveraged (e.g. material loss due to automated decision-making). These are, however, not comparable as they occur in different situations and with different respective controls, and so are characterized by a separate probability distribution. The chance of those two PDFs completely coinciding is rather small.
- **Inappropriate statistical model**—sometimes the analyst may leverage a model that is not suitable to the problem at hand. While few modelers will make obvious mistakes such as using classification models for regression problems, much more may fall victim to utilizing a model with unmet assumptions or over-interpreting the results thereof. As an apt example one can look at Value at Risk calculations during the financial crisis of 2007–2009 where intelligent organizations leveraged sophisticated modeling not realizing that the data used was only an incomplete sample and thus VaR metrics are underestimated. An additional realization came when professionals and regulators realized that for those purposes a model of extreme risks such as the Expected Tail Loss may be more appropriate, and thus the exodus away from VaR and into ETL began.
- **Misplaced causality**—the statistical algorithm is a poor substitute for genuine subject matter knowledge and by itself does not guarantee a causal link between variables. Should experts postulate an erroneous causality link the model results will reflect the error and will thus yield useless, and even harmful, estimates of effects, probabilities and outcomes. This problems is particularly exacerbated in complex environments that are characterized by high dynamics and limited knowledge.
- **Cognitive biases and frame blindness**—one of the strong points of statistical methods is that they pride themselves on being able to remove human biases and reach their result in an objective, deterministic and reproducible ways. This has been touted as one of the main reasons why even simple models outperform human judgment as early as the first showdowns between the two.[41] However, human bias may still have an effect on results as biased humans select the method, curate and process the input data, and finally interpret results. In a similar vein, frame blindness may be an issue as experts prefer certain type of modeling or

[41] Dawes [13].

even statistical paradigm (e.g. frequentist vs. Bayesian) that in inappropriate for the issue at hand. The analyst must therefore take significant care to keep an open mind as to the options when leveraging statistical modeling and the alternative interpretations of resulting estimates.

In short, the use of statistical methods has the significant benefit of formalizing intuition and grounding it into empirical reality. The sparsity of data means that often rare events cannot be estimated with precision and that a lot of inference has to lean strongly on extrapolating identified trends outside the range of observable data. Constructing statistical distributions is also challenging especially when one is unable to observe or verify the existence of extreme outliers. A remedy to this is for the analyst to remain conservative in the estimates and to leverage fat tails distributions that can account for this eventuality. While the usual *modus operandi* has been to assume the Normal distribution unless proven otherwise by a test like the Jacque-Berra one, the risk manager may be wiser to choose instead a fat-tailed distribution as a fail-safe default option.

5.3.2 Judgment-Adjusted Forecast

Following the insights gleaned from results such as Eq. (5.9), many an analyst has asked the question of whether purely statistical estimates may be improved by enhancing it with expert judgment. The rationale behind such a course of action is intuitive: rare events or effects may not yet be included in existing data but are conceivable by human experts. Thus, adjusting the purely statistical result with this knowledge may lead to improvement. Such procedures are employed both in research and in practice and in some isolated cases are able to increase accuracy.[42] On the whole, however, results are rather mixed.[43] Combined expert forecasts are still able to improve on combined model forecasts, which is mainly due to the fact that they encapsulate knowledge not visible in data but are still able to diminish the importance of cognitive biases and idiosyncratic factors through the process of aggregation.

In case of sparsity in the reference class as when projecting rare risks, integrating expert knowledge may outperform statistical methods, particularly individual algorithms. The analyst must still take heed to focus the estimation exercise on a relevant reference class. The major problems in leveraging judgment-adjusted forecasting seem to be the proliferation of misplaced causality and cognitive biases. The former group of issues arrives as the expert who is called upon to adjust the model estimate may have preconceived notions on causal links that are then imposed on the data. The great benefit of statistical modeling is that it reins in preconceptions and eliminates biases. Therefore, great care must be taken so that adjustment does not follow a wrong idea of causal links that then skews estimates. Cognitive biases that are

[42] Goodwin and Wright [16].

[43] Van Dijk and Franses [49].

typical of pure expert judgment may also pose a challenge in the adjustment process. They are supposedly lessened as the statistical forecast acts as a natural anchor for the expert adjustment, but even careful evaluations may fall prey to the availability heuristic, representativeness, or over-confidence bias.

The nature of intangible events or activities in the digital economy, as well as the fact that some tangible and measurable events are, in fact, not measured, makes it necessary to resort to subject matter experts. Initially generating a statistical forecast may be construed as a way to provide a useful anchor or even an analogy which experts can leverage. The adjustment process is still far away from straightforward and specific methods and procedures need to be followed.[44] At any rate, one is left wondering whether it makes sense to adjust a statistical forecast if data is plentiful or to generate one at all if data is too sparse. In the former case adjustment is hardly needed, and in the latter—human judgment alone may suffice.

5.4 Prediction Markets

In 2008 a group of highly distinguished social scientists, headed by Nobel-prize-winning economist Kenneth Arrow published a startling opinion piece in the journal *Science*.[45] In it they recognized the emergence of a new mechanism for eliciting forecasts on uncertain socio-economic events that sometimes gave extremely accurate predictions for fields as diverse as politics, business, and sports. This mechanism was the up-and-coming prediction markets, and the intellectuals called upon government to remove the regulatory impediments to their fast growth in view of the potential benefits they can engender.

Prediction markets are, in short, markets that are primarily run in order to use the aggregated information to make a forecast about a future event.[46] In them market participants trade contracts, and the contract payoff is dependent upon the realization of a certain event of interest. Thus, the resulting prices of those contracts (instruments) can be interpreted as reflections of the probability that a given event materializes. For example, if a contract with a payoff of EUR 1 that a certain politician wins the election trades for EUR 0.68, then one can infer that a frictionless market assigns a probability of 68% for this triumph.

One of the key theoretical questions in dealing with prediction markets is whether the actual price reflects investor beliefs in a comprehensive and accurate way. Wolfers and Zitzewitz[47] propose a simple model to study this. They begin by assuming that all traders have a constant relative risk-aversion (CRRA) utility function with risk exponent $\gamma = 1$, thus yielding a log utility of the following form:

---

[44] Franses [15].

[45] Arrow et al. [4].

[46] Berg et al. [9].

[47] Wolfers and Zitzewitz [53].

$$u(y) = \ln y \tag{5.10}$$

Traders in the prediction market have heterogenous beliefs and each trader *j*'s belief that the event of interest materializes is denoted with q_j. These have a specific statistical distribution *F(q)*, and are orthogonal to the level of wealth, *w*. Classically, investors are assumed to be rational price-takers who seek to maximize their expected utility EU_j by trading contracts x_j with a payoff of 1 and a price of π. In short, traders maximize the following equation:

$$\max EU_j = \max\{q_j \log[w + x_j(1-\pi)] + (1-q_j)\log[w - x_j\pi]\} \tag{5.11}$$

The solution to this gives the optimal number of contracts traded x^*:

$$x^* = w\frac{q_j - \pi}{\pi(1-\pi)} \tag{5.12}$$

Equation (5.12) gives specific characteristics about market demand in this stylized prediction market. First, the demand of an individual investor is zero as the price is equal to the belief of this investor. If the investor believes that the event is more likely that its price ($q_j > \pi$), then this investor will want to buy the contract. In the opposite case ($q_j < \pi$), this individual is a net supplier. This leads one to an interesting corollary. Should a prediction market be populated only by rational fully-informed traders that form beliefs in a similar way, then the price will reflect fundamentals ($q_j \approx \pi$) and no one will be willing to trade, as the profit incentive dissipates. Thus, for trade to occur, the market needs heterogenous agents and beliefs and it may be important to attract less well-informed traders to ensure sufficient levels of exchange and liquidity.

Second, for investors with a given wealth level *w*, demand for the contract increases linearly with their beliefs, and decreases with the risk. Greater risks, particularly as the price approaches 0.5 lead to a lower level of demand. Third, the demand is homothetic, meaning that it increases proportionately with an increase in wealth. Essentially, investors that can afford to lose more, will tend to bet more in the prediction market. This simple model also yields a unique level of demand, making it both intuitive and tractable. As in any market, equilibrium occurs under the condition that demands equals supply, or:

$$\int_{-\infty}^{\pi} w\frac{q - \pi}{\pi(1-\pi)} f(q)dq = \int_{\pi}^{\infty} w\frac{\pi - q}{\pi(1-\pi)} f(q)dq \tag{5.13}$$

Assuming the independence of investor beliefs (*q*) and means (*w*), then this implies the following equilibrium condition:

$$\frac{w}{\pi(1-\pi)}\int_{-\infty}^{\pi} (q-\pi)f(q)dq = \frac{w}{\pi(1-\pi)}\int_{\pi}^{\infty} (\pi - q)f(q)dq \tag{5.14}$$

From Eq. (5.14) it follows that the price of the contract reflects the mean belief of the traders, or:

$$\pi = \int_{-\infty}^{\infty} (q - \pi) f(q) dq = \overline{q} \tag{5.15}$$

The result in Eq. (5.15) is a rather strong one. It shows that pricing in prediction markets accurately reflects the estimates of their participants. An important caveat here merits attention. This result is predicated on the use of logarithmic utility function. However, when alternating between a set of plausible alternative utility specifications such as alternative CRRA specifications, constant absolute risk aversion (CARA) specifications, quadratic utility and hyperbolic absolute risk aversion (HARA), the results largely carry over. Thus, for a wide range of plausible alternatives, market prices reflect investor beliefs. This is also supported by a range of empirical results, showing that market prices tend to be close to mean beliefs and to actual outcomes. Furthermore, prices are fairly efficient estimators of realizations. All this gives credence in using prediction markets for generating forecasts for the thorny problem of quantifying rare risks.

The prediction market can be designed in a variety of ways, whereby the payoff is linked to numerous features of the event. The four most popular approaches are as follows[48]:

- *Instrument linked to binary outcome*—it this case the contracts reflect the occurrence of an event such as whether the stock market goes up or the incumbent wins an election. The aggregated instrument payoff reflects the probability of occurrence.
- *Instrument linked to continuous outcome*—this contract type is linked to a relevant continuous variable such as the percentage growth of the stock market or the proportion of votes, received by the incumbent. Now the contract reveals the market expectation of the mean of the probability distribution.
- *Instrument linked to combination between binary and continuous outcome*—if needed, a combination of predictions may be included in a given contract. Then, its resulting value will reflect the median expectation of the market.
- *Winner-takes-all family of contracts*—in this case a set of different contracts where each reflects a different outcome may be traded. Their totality enables the analyst to infer the complete probability distribution of the event of interest. Working with a set of contracts pertaining to correlated events also enables one to calculate not merely the unconditional probabilities but also the conditional ones of realizations given other events. Those inferences generally take recourse to the Bayesian theorem or more general Bayesian methods.

At any rate, the importance of an appropriate design of the prediction market is crucial for the elicitation of the forecast of the variable of interest. In the case of rare events this is even more so as participants are unlikely to have been exposed

[48] Wolfers and Zitzewitz [53].

to those and their correct calibration is suspect. What is more, the analyst may only do corrections on resulting values with very little structured information on which ones are closer to reality. Indeed, concerns have been raised on some potential issues with using prediction markets, most notably the willingness of traders to trade, the contractability of relevant outcomes, the possibility for manipulation, the expert calibration, and the misplaced causality issues.[49] Market design is called upon to rectify all of them. All in all, this seems to be successful as research has shown that prediction markets tend to be efficient according to the Efficient Market Hypothesis, displaying at least weak efficiency.[50] This can be further enhanced by adjusting the estimates generated with either time series on prediction, or with additional poll questions,[51] thus further improving the quality of estimates.

Even initial research on prediction markets did not fail to point at their obvious utility in being used as a decision support mechanism.[52] This is hardly surprising as they provide a formal aggregation of the wisdom of the crowds achieved through an incentive-enabled market environment. Some initial results even point to the fact that prediction markets are able to provide superior forecast performance when compared to the averaging of expert opinions within a forecasting poll.[53] It is thus reasonable to employ prediction markets for the elicitation and aggregation of risk estimates but one important difference from human judgment persists. Prediction markets are still markets. This means that they are subject to price dynamics, irrational investor behavior in search of profits, and possible market imperfections.

It is hardly surprising then that prediction markets operate like, and exhibit many of the characteristics of, emerging financial markets.[54] Most notably, price changes in prediction markets are relatively more likely to be negative than positive but when the price changes in the positive direction, this trend is much stronger. This leads to a marked right skew of the overall distributions in the market. Thus, while falls are more likely, spectacular growth dominates them. This is also reflected in the fat tails of the contract return's statistical distribution. Interestingly, research on 3385 betting markets shows that either of those fat tails, when looked in separation, follows a power law distribution. Prediction markets are also found to have some long-term term memory but little autocorrelation, further hinting at their efficiency.

While a relatively new method for information elicitation, prediction markets have shown significant promise. They are able to generate useful forecasts but sometimes, as any market, they can fail spectacularly. In extreme cases such as in a closely contested election, they lag behind even unsophisticated instruments such as the

[49] Wolfers and Zitzewitz [53].

[50] Angelini and De Angelis [1].

[51] Dai et al. [12].

[52] Berg and Rietz [8].

[53] Atanasov et al. [5].

[54] Restochhi et al. [42].

opinion poll.[55] Thus, prediction markets are best considered as yet another instrument for aggregating individual estimates that can be leveraged for risk management purposes as needed but are in need of further investigation before they can wholeheartedly replace currently existing methods.

5.5 Monte Carlo Methods and Scenarios

Running simulations that leverage Monte Carlo methods is a key tool for the risk analyst.[56] The name Monte Carlo itself is enticing but in the context of risk analysis it has little to do with glitzy casinos and fortunes being made or lost. Still, the logic behind it is quite straightforward. Starting from the premise that actual available data is necessarily a sample of possible realizations, the analysis proceeds to generate a large number of simulated alternative scenarios that give an idea of many more possibilities. The logic behind this process is visually presented in Fig. 5.3, and proceeds along the following steps:

1. **Identify variables of interest and their connections**—a first natural step is to identify the target variable under investigation (such as losses from operational or financial risks) and how it depends on other features. This dependence may be a statistical one as summarized in a market beta or some correlation (e.g., dependence of stock market returns on economic growth), or an identity equation (e.g., profit being equal to revenue minus cost). This preliminary stage may involve both efforts in data mining that uncover unexpected patterns, as well as expert evaluations that complement and enhance domain understanding.
2. **Investigate the statistical properties of their distribution**—the second step is to obtain and curate sufficient data of adequate quality that pertains to the variable under scrutiny. Detailed quantitative data of risk realizations would be most suitable for this and can sometimes be obtained from transaction systems, process-aware information systems, or different event logs. Those may need to be further processed, normalized, and possibly enriched. In some cases, such data are simply not available, and the analyst will have to depend on information elicitation from experts.[57] The key point here is to obtain evaluations of the parameters that allow the reconstruction of the loss distribution function, the event frequency distribution function or the resulting probability density function for the expected losses and gains. The latter can be reconstructed by asking questions such as what is the range of outcomes that a realization will fall into with a given level of certainty (e.g. 95%). Knowing that the loss of a given action will be between 1 and 5 million with a 95% certainty, and assuming a normal distribution, the analyst can easily reconstruct the whole PDF.

---

55 Mann [32].

56 See e.g. Brandimarte [10].

57 Baptistucci et al. [7].

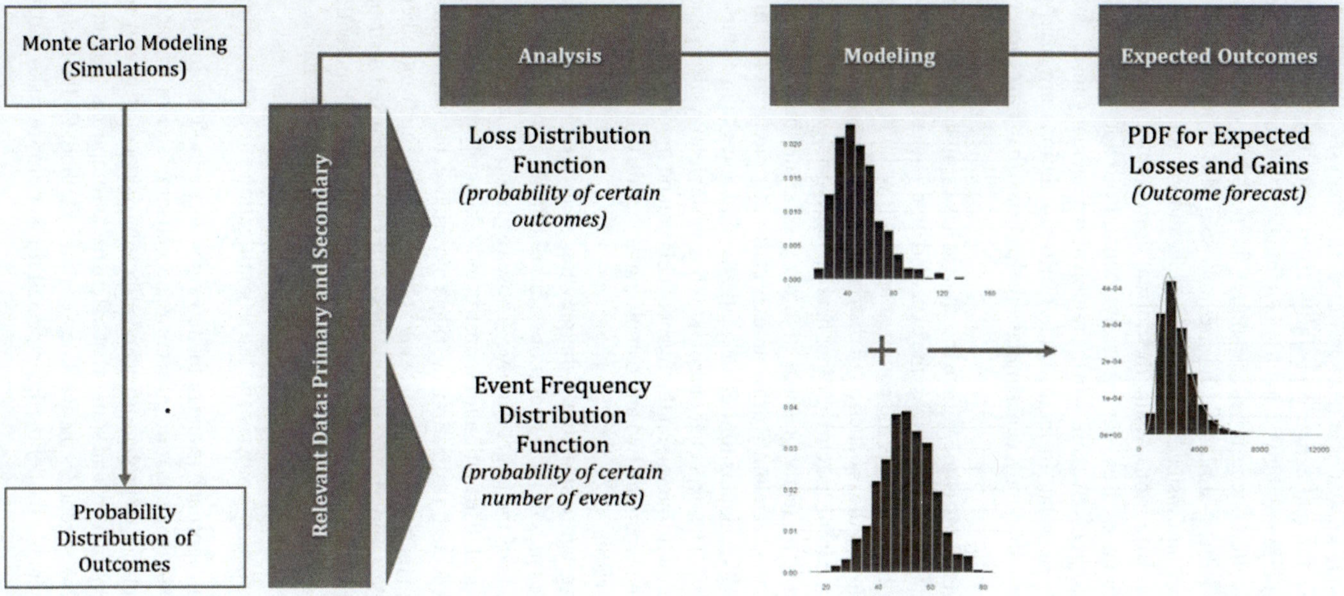

Fig. 5.3 Monte Carlo simulations logic overview, source: Esterhuysen et al., [14]

3. **Simulate probability value of realization**—once the type of distribution and its key parameters are known or at least assumed, the analysis proceeds by drawing a random value for realization probability from its respective distribution. A Gaussian distribution will give a natural benchmark, but the analyst may choose to rely on a fat-tailed one as it better captures the possibility of very rare risks.
4. **Simulate loss value**—similar to the previous step, the expected loss may not be deterministic, and instead be modeled as a random number drawn from its own statistical distribution. Again, the normal or the uniform distributions may be useful benchmark cases, but the analyst should provide for the long tail of realizations. It is precisely there—in the region of spectacular gain or catastrophic losses that rare risks lie. In fact, Monte Carlo simulations are used precisely because of this—that they can simulate realizations that have never actually occurred in real life.
5. **Estimate expected outcome**—once the analyst has estimated the probability of occurrence and the associated loss, the expected outcome can be easily obtained by multiplying them. That is the common approach when the granularity of the analysis has allowed the production of distribution functions for both probability of occurrence and impacts. On the other hand, if data sparsity or deliberate choice leads to that, it is perfectly possible to directly simulate the PDF for expected outcomes, and then randomly draw an expected outcome from it. This may be necessary whenever data is limited or qualitative in nature such as the case with expert elicitation. The expected outcome of one variable may affect another one as the latter is conditional upon the former. This can be modeled by the strength of association between the two and can ultimately modify the expected outcome of the target variable(s).[58]
6. **Iterate and aggregate statistics**—the previous steps are usually repeated over numerous (thousands or more) iterations, thus obtaining a large number of simulated realizations of the target variable. Those realizations are then summarized and their statistical distribution is reconstructed. The analyst obtains a clear idea of the mean (expected) realization, the range of variance, the likelihood of certain occurrence (e.g. the probability of losing at least 10% of value) and other relevant information. Most importantly, at the edges of this reconstructed distribution one finds the rare risks lurking. Those are the unlikely events with potentially large impact. By studying the tails of the target variable PDF, the analyst obtains a feeling of the magnitude and propensity for large, unexpected risks. This can then inform the overall analysis and the risk mitigation efforts stemming from it.

The Monte Carlo approach is one of the few rigorous methods to do a quantitative simulation of events that are rare in occurrence and dramatic in impact. It relies on reconstructing plausible probabilities and effects of risk occurrence by relying on proxy data, expert estimates, and somewhat extensive assumptions. This creates both the strengths but also engenders the weaknesses of the method. Of the former, Monte Carlo simulations enable modeling the expected impact of rare risks and sensitizes

[58] See Zhang et al. [55].

different stakeholders to plausible turns of events, thus improving risk management. The building blocks of the simulation may be individually well-understood (such as failure probabilities of distinct machine components) but it may still yield emergent properties with surprising results at the aggregate level. Additionally, the method enables the analysis of simultaneous uncertainties over many points in a complex system and their interplay.[59]

On the other hand, the results obtained are only as good as the input data and simulated scenarios can be very sensitive to starting assumption such as the type of statistical distribution and its corresponding tuning parameters. This may lead to a decreased level of trust in the rigor of the analysis, and one should be careful to perform sensitivity checks to guard against this. Furthermore, the rare risk in the distribution's tail may still be perceived as insignificant due to its low likelihood and stakeholders may be tempted by the seeming numerical precision of the estimates to make a decision on the margin. However, Monte Carlo scenarios should be used as a rough guideline since the actual realizations may differ by orders of magnitude by virtue of the construction and parametrization of the simulation.

5.6 Modeling the Unexpected in the Digital Economy

Niels Bohr famously said that prediction is very difficult especially when it is about the future. This is the very issue with formally analyzing rare risks—since they occur only very infrequently, or even have not yet occurred, the analyst will have to generate plausible data points that can then be leveraged in the risk management process. The Monte Carlo simulation is a key tool for that, and it allows one to model even very complex and emergent phenomena such as those in the digital economy. The scarcity of data necessitates more assumptions compared to well-understood problem domains but the key to tractability is generating explicit and clear ones that are then subjected to sensitivity testing.

A first step would be conceptualizing the effects of the risk. From an economic and business standpoint, a natural focus of interest would be revenue dynamics and how they could possibly be impacted by uncertainties in the complex environment. To this end we select to model the growth of revenue in all data processing sectors (NACE code J63) by investigating its drivers and formally modeling their dynamics. Data is taken for 28 European Union countries in the period from Q1.2000 to Q4.2020. Chapter 4 has already modeled their interactions via a Bayesian Belief Network and the simulation exercise follows the same logic and utilizes the same variables. However, all chain-linked indices are now transformed to percentage growth, so that results from the Monte Carlo simulation are more intuitive to interpret.

The approach to constructing the Monte Carlo model follows an adapted version of the one presented above, entailing the following steps:

[59] Rezaie et al. [43].

1. **Select target variable** of interest that is relevant to the risk analytic exercise and can provide valuable insight from academic or practical standpoint. The target variable here in information services revenue (termed J63 Turnover).
2. **Identify the drivers** of the target variable movement and estimate their effect (such as betas, elasticities, or other relevant metrics). The identification of drivers may proceed through leveraging expert knowledge, or a combination of both. The current model uses the structure automatically generated by the Bayesian Belief Network in Chap. 4.
3. **Model each individual driver** and select an appropriate statistical distribution that characterizes its dynamics well by allowing for unexpected risks to occur. This may include departing from the usual assumptions of normality and using a fat-tailed distribution instead. For its versatility, the Student T distribution is used as it allows modeling both the normal distribution and variations of fat-tailed ones.
4. **Generate simulated driver values** using the defined statistical distribution over many periods to allow for the materialization of rare events at the tails of the distribution. The number of iterations selected is $N = 10{,}000$.
5. **Estimate target variable** over each simulation iteration. At each round, this variable is defined by the values of the structural drivers and the strength of their effects as measured by the beta coefficients in a regression equation.
6. **Construct and investigate the statistical distribution** of the target variable using data points over all the iterations. We also look at its overall shape and form, test for normality, estimation of distribution moments, and others.
7. **Perform sensitivity testing** by checking assumptions, varying distribution parameters, re-estimating the simulation, and comparing results. This step is crucial as the Monte Carlo may generate very different outcomes depending on slight assumption changes or even on little variations in its random components. A particular focus here is to understand how the structure of unexpected external shocks (rare risks in the drivers) eventually affects the target variable.
8. **Draw conclusions for the risk analysis** based on the insights from the simulated system. Apart from generating rigorous results, the risk analysis will have to take into account the level of uncertainty inherent in generating new data out of thin air.

Following those steps, a Monte Carlo simulation for the digital economy is generated that enables one to develop a grasp of the possibility of rare but highly impactful occurrences.

5.6.1 Overall Model Structure

Turnover growth in sector J63 (Information service activities) across Europe has been quite uneven in the different counties in the continent. While some have witnessed a steady trend growth, others exhibited explosive trajectories. This in turn leads

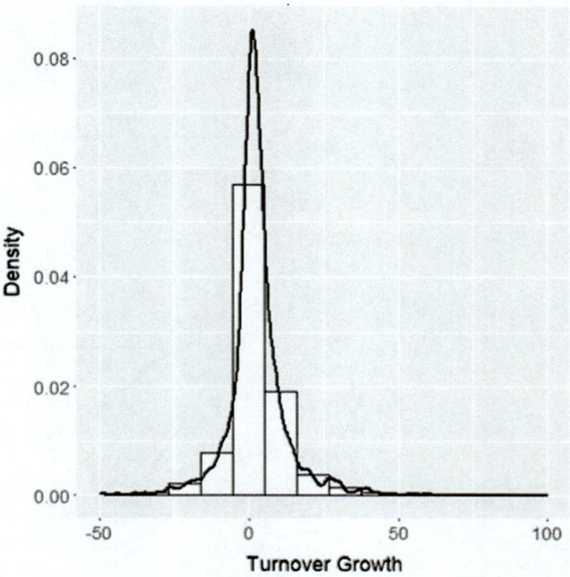

Fig. 5.4 Distribution of turnover growth in sector J63

to a highly skewed non-normal distribution of turnover data (see Fig. 5.4). While most data points are centered in the middle around the mean of 3% growth, the distribution has a markedly long tail, with outliers on the positive side. There is also quite a significant level of volatility in the turnover throughout the two decades under study as evidenced by a standard deviation of 12.5%. The Monte Carlo simulation will not model J63 directly, but rather its drivers and will reconstruct the risk profile based on their values. However, a useful simulation should be able to mimic basic characteristics of actual turnover data, while allowing the analyst to see possible realizations that have not yet materialized in the sample.

Understanding the problem domain is crucial and the Monte Carlo model builds upon the insights generated from the Bayesian Belief Network in Chap. 4. In essence, it shows that nine macroeconomic and sectoral variables tend to influence revenue in sector J63. Those are: inflation (HICP), compensation of employees, interest rate, export of goods, household consumption, economic growth, import of goods, import of services, and labor costs in IT (code J60).

The strength of their effects (betas) is recalculated as data is transformed from CLI to growth rates. Those betas are obtained within a multiple regression framework and thus have the interpretation of time-invariant non-probabilistic effects. On the one hand, this makes for a more straightforward interpretation, while on the other, those estimates are quantitatively very close to the ones obtained within a Bayesian network model. The chain and strength of influence is presented graphically in Fig. 5.5. The distribution of each driver is then formally modeled. To allow for more versatility, the assumption of normality is dropped and each variable is instead assumed to follow a Student T distribution that can accommodate both Gaussian and fat-tailed data, depending on the degrees of freedom parameter.

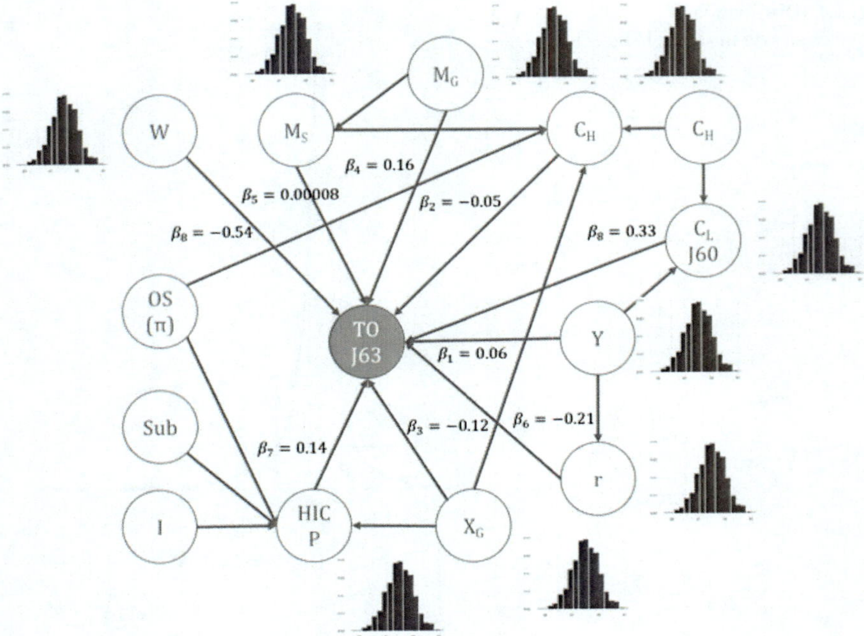

Fig. 5.5 Recalculated risk betas of J63 turnover with respect to simulated variables

5.6.2 *Key Results*

The descriptive statistics of all the drivers of J63 revenue are presented in Table 5.1. Most of the variables under study are relatively stable, with only a few have a more pronounced volatility. Among those are the growth of the employee compensation, the IT labor cost growth, and the overall economic growth. Revenue in J63 is by far the most volatile among the aggregates under investigation with a standard deviation of more than four times the mean. All the variables have a Jarque–Bera test statistic that is sufficient to reject the null of normal distribution at a very high level of statistical significance ($p < 0.0005$). To accommodate this effect, all the drivers are assumed to follow a variant of the Student T distribution. The exact parameters of each distribution are automatically calculated using a maximum likelihood estimator[60] (MLE) and shown in Table 5.1. For only two of the variables—import and export of goods, the degrees of freedom parameter is sufficiently large to approximate a gaussian distribution, while the rest are mostly fat-tailed ones.

Using the estimated parameters, nine different distributions—one for each driver—are reconstructed. A random value is drawn from each, proxying a plausible realization of this driver in real life. Then the betas are used to calculate the

[60] Ripley et al. [50].

Table 5.1 Key descriptives and fitted parameters for Student T distribution for selected variables

Variable	Mean	SD	Min	Max	Jarque Bera, χ^2	T dist. mean	T dist. scale	Degrees of freedom
Inflation (HICP)	0.52	1.07	−4.91	6.55	1951.07	0.44	0.66	2.94
Compensation of employees	0.12	1.97	−13.95	14.44	7951.93	0.05	0.96	2.14
Interest rate	3.74	2.60	−0.54	25.40	6046.45	3.61	2.08	6.49
Export of goods	36.54	17.54	7.20	86.00	119.80	36.54	17.53	>100
Household consumption	54.96	8.56	23.20	71.50	238.12	55.36	7.90	13.78
Economic growth	0.49	2.17	−18.75	17.05	43,218.4	0.61	0.63	1.60
Import of goods	41.14	15.40	17.20	85.40	110.96	41.14	15.39	>100
Import of services	18.17	26.23	4.10	151.80	10,187.2	8.90	2.74	1.00
Labor costs in IT	1.03	2.98	−13.64	21.04	2643.26	0.85	1.51	2.07
J63 Turnover	3.00	12.47	−48.96	240.18	550,919	1.97	4.17	1.60

effects of each driver and to estimate a final resulting value of the target variable—J63 Turnover. This process is repeated 10,000 times to obtain values for very rare events—indeed, using quarterly data over 10,000 periods means generating simulated time series for 2,500 years! This then paints a somewhat deterministic picture of the world, while there may be numerous random shocks affecting the drivers under investigation. While they may not change the overall shape of the distribution, they have strong effect over a single period. A constellation of such random shocks in the same direction (positive or negative) may have immense effects on the target variable in this period and be alone responsible for generating the rare risk. Colloquially, using such random shocks allows the modeling of "perfect storms".

To investigate their effects and the sensitivity of the Monte Carlo simulation to alternative shock assumption, four different versions of the model are simulated:

- **Simulation with No Shocks**—the assumption here is that there is a deterministic effect of the (randomly generated) driver values on the target dependent variable. This entails that the betas remain time invariant, and also that the drivers themselves are not subjected to sizable external influences. While not particularly realistic, the no-shock simulation may serve as a useful baseline.
- **Simulation with Normally Distributed (Gaussian) Shocks**—in this case the assumption is that variables are subject to normally distributed exogenous hits. The residuals of the regression equation with turnover as the dependent variable

and the drivers as the independent ones may give an idea of the distribution of shocks. Residuals are thus estimated, and the normal distribution parameters are fit on them, using a maximum likelihood estimator. Then, we reconstruct the shock distributions and draw random values from them, adding those to the drivers in the given period. The mean of the shocks is zero, and thus the overall driver distribution is changed little but for more randomness.

- **Simulation with Exponential Shocks**—starting from the premise that many events in the digital world are exponential in nature, then it would be natural to model the unexpected shocks as following an exponential distribution. In practice, this would mean that many events will be small-scale ones, but the impact will be dominated by very few, very large occurrences. From a practical standpoint, exponential distribution is fitted on, and generates only, positive values. The analyst may overcome this by assigning a random multiple in front of the generated number that takes the value of +1 or −1 with a given probability (e.g., 0.5). Again, regression residuals are used to reconstruct the exponential distribution and to estimate its scale value. Mean of those shocks is also zero.
- **Simulation with Long-tailed Shocks**—looking at Fig. 5.4 shows that the distribution of turnover features a long tail and visually approximates the lognormal distribution. This effect may be driven by a lognormal shape of the shocks that hit revenue drivers. The fourth simulation thus fits the lognormal distribution parameters and randomly draws exogenous shocks from it. As with all simulations above, the mean of those shocks is zero. They still serve to expand the realization space so that the analyst may obtain a useful idea of possible but extremely rare and unlikely realizations of the variables under study.

Histograms of the actual data (N = 1,789) and the three simulations (N = 10,000 each) with shocks are presented visually in Fig. 5.6. The figure clearly shows how the Monte Carlo models expand and enrich the space of existing data with newly generated realizations. Those now feature much more events in the extreme ends of the distribution tails and are thus able to include rare risks that may have yet to realize. The structure of shocks also affects the way that existing data is expanded. The Gaussian shock structure serves to flatten the distribution with many scenarios clustering around the mean of 3% but also providing a relatively high probability of outliers. Exponential shocks lead to a distribution where outliers are extreme but not very likely. Finally, the lognormal-shocks structure features extreme events with relatively higher likelihood. The Monte Carlo simulation without shocks (not shown) is by far the most trivial one—it has values that neatly cluster around the distribution mean and practically no events at the tails, thus making it an unsuitable candidate for modeling the rare and unexpected black swan.

The numeric characteristics of the actual data and all simulated scenarios are shown in Table 5.2. All four Monte Carlo simulations are able to reproduce the mean growth rate of J63 revenue with significant accuracy. With a baseline mean of 3% in the actual data, the mean values in the simulations vary in the range of 2.97–3.22%. The volatility of actual data is another matter. The simulation with no shocks exhibits very limited fluctuations with a standard deviation of 3.81%, against an actual

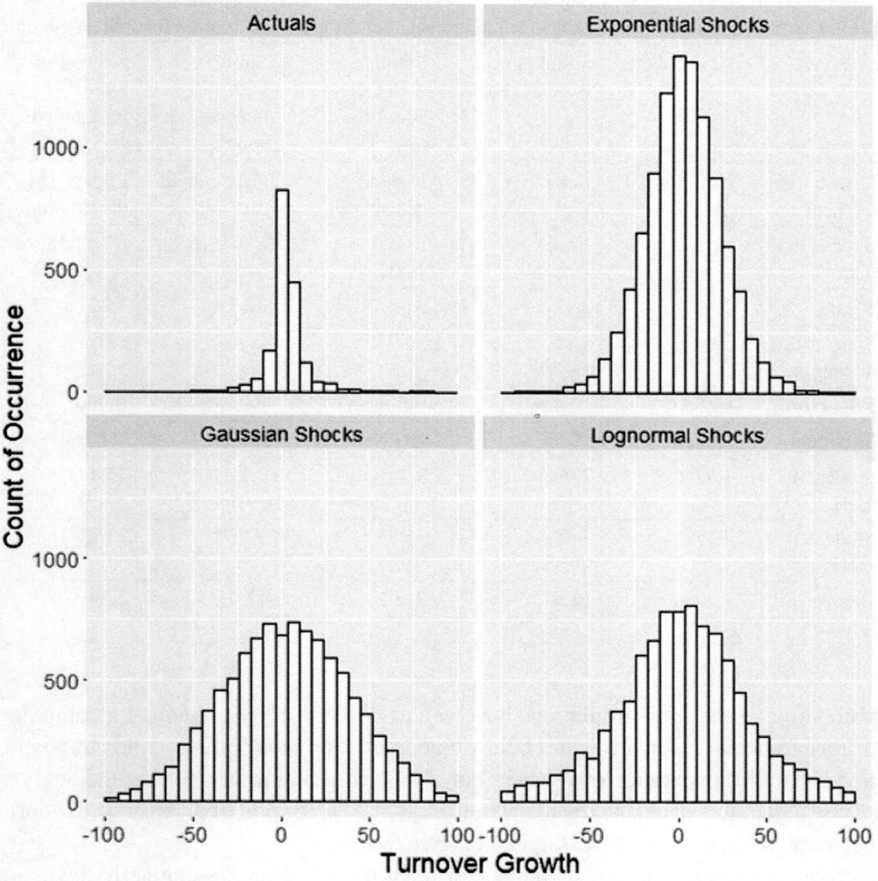

Fig. 5.6 Histograms of Monte Carlo simulations under different assumption for shock structure

baseline of 12.47%. The simulations with normally and exponentially distributed errors show much higher volatility rates of 2 to 3 times the baseline. The long-tailed shocks simulation has the largest fluctuations of them all with a standard deviation of almost 53%. The difference in the standard deviations between simulations and actuals are not necessarily a bad thing—they show unexpected volatility and are driven by events in the tails of the distribution. It is precisely the presence of those rare risks that make the Monte Carlo simulation a useful tool.

The maximum value of actual revenue growth is 240% in the real data, and this is hardly captured by most simulations. The ones with exponential and gaussian shocks only reach levels of half that magnitude. The simulation with lognormal errors fares better—it has an extreme value of 661%—a rare and unexpected event, indeed. Overall, simulations may be underestimating risks on the downside with three of the Monte Carlo models showing lower value that the actual ones. This is exactly reversed with black swans—the actual worst-case scenario is a revenue drop of about 49%,

Table 5.2 Key statistic of Monte Carlo simulations under different assumption for shock structure

Feature	Actual data	Simulation without shocks	Simulation with Gaussian shocks	Simulation with exponential shocks	Simulation with long-tailed shocks
Distribution of shocks	Non-normal	None	Normal distribution	Exponential distribution	Lognormal distribution
Shocks PDF parameters	Unknown	None	$\mu = 0, \sigma =$ 12.32	$\lambda = 0.15$	$\mu = 1.18, \sigma$ $= 1.33$
Mean	3.00	2.97	3.13	3.13	3.22
Standard deviation	12.47	3.81	36.98	21.64	52.55
Minimum	−48.96	−34.89	Below −100	Below −100	Below −100
Maximum	240.18	48.58	137.70	99.22	660.99
Student T mean	1.97	2.98	3.14	3.12	3.34
Student T scale	4.17	3.45	36.54	19.77	23.23
Degrees of freedom	1.60	12.07	81.92	12.01	2.09

while three of the four simulations have values below −100%, meaning a complete disappearance of revenue. Values below that do have an interpretation, indicating not merely the disappearance of revenue but also accumulation of debt but the analyst may still decide to truncate the distribution at −100% for practical reasons. In short, the simulations tend to overestimate negative events.

Those effects are in play as the sector of information processing activities is in rapid development and with strong fundamental demand. This translates into the fact that growth is positively skewed, and also that favourable rare occurrences are more likely than negative ones. This is not captured very well by a symmetric distribution as is used here. However, as the sector matures, and becomes strongly integrated within the overall economy, it is likely that negative rare events will also become somewhat more likely. The analyst may thus choose to retain a single distribution for the exogenous shocks or, as an alternative, model the positive and negative ones separately.

Investigation of the data in Table 5.2 shows that standard Monte Carlo simulations without a rich structure of exogenous errors, or those that assume a set of normally distributed shocks may not be suitable for modeling phenomena in the digital world. Those two simulations are least able to approximate real data and are limited in modeling extreme risks, particularly on the positive side of the distribution. Due to the exponential nature of many digital events, distributions with long tails such as the exponential one and the lognormal one seem better suited not only to capture some stylized facts of actual data but also to suitably expand the space of realizations so that rare and unlikely occurrences are generated. The analyst is thus advised to consider switching away from the Normal distribution and into distributions that are more suitable for modeling extreme digital events that are loosely or not at all bounded by physical limits. Finally, the results from the simulations enable the construction of a separate distribution function of the target variable (J63 Turnover) with MLE-fitted parameters. The ones for the Student T distribution are also shown in Table 5.2 and can be used to reconstruct the turnover distribution and calculate standard risk metrics on it such as the Value at Risk or the Expected Tail Loss.

5.7 Conclusion

Analyzing rare risks is the challenge it is because it requires a leap of faith that stretches the bounds of imagination. The analyst here deals with extremely unlikely occurrences for which limited, or no data is available. Thus, some entity has to imagine what may happen—this may be the proverbial domain expert, a wise crowd, a prediction market, or even a random number generator. Despite the route taken, the fact remains that risk analysis has to proceed with very limited data on empirical probability of these rare black swans. In fact, the task could be so enormous as to tempt the analyst to leave it. The main argument against this choice is that rare events tend to dominate the risk exposure of organizations and individuals and their consequences may be catastrophic indeed. From nuclear power plant explosions through global financial meltdowns, we do know that even very rare events do occur and have to guard against them accordingly. On the bright side, those risks may have a positive impact, such as the rise of the internet, that can revolutionize businesses for good.

The approach to analyzing rare risks in the digital age proceeds along the familiar line of identifying them, estimating their probability, impact, and possibly correlations, and then using this insight to form a structured risk management decision. The analytic part may take recourse to many sources of (imperfect) information. First, experts may be enlisted to help. This may either be done in individual sessions or in group formats such as the workshop where different pieces of knowledge combine to expand the view of what plausible risks may occur. Research has shown that a key determinant of the quality of elicited information is the structure of elicitation. Experts can very well fall prey to biases such as tunnel vision and group thinking, and overconfidently stick to their erroneous estimates due to confirmation bias. Thus, methods such as the structured analogy whereby more information is given, or iterative approaches such as the Delphi method, may be helpful.

Second, statistical methods may be used that leverage the scarce existing data. As risk is always a phenomenon of the future, forecasting models are the most common choice. The analyst may take recourse to pure time series ones whereby only the structure of the data realizations over time is used, leveraging the predictive power of the lag and error structure. Alternatively, structural or mixed models may be employed so that process drivers impact the forecasted target variable of interest. The main issue with statistics is that the forecast generated is only as good as the data ingested. If the dataset contains no realizations of the rare event, then it cannot possibly be forecasted! Furthermore, if the dataset underrepresents rare risks, they will be dominated by standard occurrences and smoothed over, thus all but disappearing. This possibility is even greater should the analysis focus on novel and emergent phenomena such as those in the digital sectors of the economy.

Third, a formalized way to elicit the wisdom of a large group of experts (and probably amateurs) is to structure a prediction market. It functions just like an ordinary market, but the goods exchanged are predictions about the realization of future (possibly rare) events. The pricing mechanism can be structured in such a way as to reflect the chance of a given even occurring, thus enabling the analyst to generate a probabilistic forecast of some risk of interest. While very promising, most prediction markets are still in their infancy and are not particularly deep—transactions may be few, and nothing guarantees that enough competent parties will be interested in participating. Moreover, the variety of forecasts traded may be limited and not cover particular risks that the analyst is interested in. To ensure uptake, prediction markets tend to cover popular events of global interest (e.g. presidential elections and world championships winners) and specific risks (that your organization may be subject to a devastating cyber-attack). Some have proposed to overcome this by instituting internal prediction markets for organizations and groups of interest, but their uptake, utility and cost–benefit ratio remain a challenging problem of its own.

Fourth, a possible way to unleash one's imagination in a rigorous manner is to create a Monte Carlo simulation that is based on actual empirical data but expands the possibilities of rare risk occurrence by assuming certain behavior of the variables. This may be introduced by leveraging fat-tailed or exponential distributions or including exogenous shock of different magnitude. As a result, the simulation is grounded in reality but provides a clearer idea of a set of plausible realizations. Running its course over a large number of periods, this Monte Carlo models eventually generates artificial data worth thousands of years. In it, the analyst may discern the rare risk and estimate its probability and impact. The simulation is even more robust as the target variable is not directly simulated but instead the analyst uses a set of drivers whose behavior is clearer and thus easier to model. Monte Carlo simulations generate all the necessary data needed for rigorous analysis of risk for even highly uncertain and unlikely events. On the flipside, one should never forget that this data is essentially all made up and provide the necessary sanity checks, reliability, and sensitivity analyses.

Predicting unpredictable risks is a task for the disciplined imagination. The analysis may make use of a wealth of methods and approaches to obtain ranges for plausible outcomes but in the end of the day decisions have to be made what a plausible range is. Stakeholders may be too conservative in this respect, grounding their deliberations in personal experience. The search for rare risks must necessarily go beyond immediate experience and into the realms of conjecture. The risk analyst has to structure the process and enable deliberations to happen within a largely expanded outcome space. Due to vested interests, entrenched attitudes, culture and habit, organizations and individuals may be unable or unwilling to embrace this. The greatest challenge lies in overcoming those.

References

1. Angelini, G., De Angelis, L.: Efficiency of online football betting markets. Int. J. Forecast. **35**(2), 712–721 (2019)
2. Armstrong, J.S.: How to make better forecasts and decisions: Avoid face-to-face meetings. Foresight: Int. J. Appl. Forecast. **5**, 3–15 (2006)
3. Aromí, J.D.: Medium term growth forecasts: experts vs. simple models. Int. J. Forecast. **35**(3), 1085–1099 (2019)
4. Arrow, K.J., Forsythe, R., Gorham, M., Hahn, R., Hanson, R., Ledyard, J.O., Neumann, G.R.: The promise of prediction markets. Science **320**(5878), 877 (2008)
5. Atanasov, P., Rescober, P., Stone, E., Swift, S.A., Servan-Schreiber, E., Tetlock, P., Mellers, B.: Distilling the wisdom of crowds: prediction markets vs. prediction polls. Manag. Sci. **63**(3), 691–706 (2017)
6. Bao, C., Wu, D., Wan, J., Li, J., Chen, J.: Comparison of different methods to design risk matrices from the perspective of applicability. Procedia Comput. Sci. **122**, 455–462 (2017)
7. Baptistucci, C.B., Pech, G., Carvalho, M.M.: Experts' engagement in risk analysis: a model merging analytic hierarchy process and Monte Carlo simulation. J. Modern Project Manag. **6**(1) (2018)
8. Berg, J.E., Rietz, T.A.: Prediction markets as decision support systems. Inf. Syst. Front. **5**(1), 79–93 (2003)
9. Berg, J., Forsythe, R., Nelson, F., Rietz, T.: Results from a dozen years of election futures markets research. Handb. Exp. Econ. Results **1**, 742–751 (2008)
10. Brandimarte, P.: Handbook in Monte Carlo Simulation: Applications in Financial Engineering, Risk Management, and Economics. Wiley (2014)
11. Chang, W., Berdini, E., Mandel, D.R., Tetlock, P.E.: Restructuring structured analytic techniques in intelligence. Intell. Natl. Secur. **33**(3), 337–356 (2018)
12. Dai, M., Jia, Y., Kou, S.: The wisdom of the crowd and prediction markets. J. Econometrics (2020). https://doi.org/10.1016/j.jeconom.2020.07.016
13. Dawes, R.M.: The robust beauty of improper linear models in decision making. Am. Psychol. **34**(7), 571 (1979)
14. Esterhuysen, J. N., Styger, P., Van Vuuren, G.: Calculating operational value-at-risk (OpVaR) in a retail bank. S Afr. Econ. Manage. Sci. **11**(1), 1-16 (2008)
15. Franses, P.H.: Expert Adjustments of Model Forecasts: Theory, Practice and Strategies for Improvement. Cambridge University Press (2014)
16. Goodwin, P., Wright, G.: The limits of forecasting methods in anticipating rare events. Technol. Forecast. Soc. Chang. **77**(3), 355–368 (2010)
17. Goossens, L.H., Cooke, R.M.: Expert judgement—calibration and combination. In: Workshop on Expert Judgment. Aix En Provence, France (2005)
18. Green, K.C., Armstrong, J.S.: Structured analogies for forecasting. Int. J. Forecast. **23**(3), 365–376 (2007)
19. Grigore, B., Peters, J., Hyde, C., Stein, K.: A comparison of two methods for expert elicitation in health technology assessments. BMC Med. Res. Methodol. **16**(1), 1–11 (2016)
20. Hastie, T., Tibshirani, R., Friedman, J.: The Elements of Statistical Learning: Data Mining, Inference, and Prediction. Springer Science & Business Media (2009)
21. Hemming, V., Armstrong, N., Burgman, M.A., Hanea, A.M.: Improving expert forecasts in reliability: application and evidence for structured elicitation protocols. Qual. Reliab. Eng. Int. **36**(2), 623–641 (2020)
22. Hertwig, R.: Tapping into the wisdom of the crowd—with confidence. Science **336**(6079), 303–304 (2012)
23. Hubbard, D.W., Drummond, D.: How to Measure Anything. Tantor Media (2011)
24. Kahneman, D.: Thinking, Fast and Slow. Macmillan (2011)
25. Koehler, D.J., Brenner, L., Griffin, D.: The calibration of expert judgment: Heuristics and biases beyond the laboratory. Heuristics Biases Psychol. Intuitive Judgment 686–715 (2002)

26. Lee, W.Y., Goodwin, P., Fildes, R., Nikolopoulos, K., Lawrence, M.: Providing support for the use of analogies in demand forecasting tasks. Int. J. Forecast. **23**(3), 377–390 (2007)
27. Lin, S.W., Bier, V.M.: A study of expert overconfidence. Reliab. Eng. Syst. Saf. **93**(5), 711–721 (2008)
28. Makridakis, S., Hibon, M.: The M3-competition: results, conclusions and implications. Int. J. Forecast. **16**(4), 451–476 (2000)
29. Makridakis, S., Hogarth, R.M., Gaba, A.: Forecasting and uncertainty in the economic and business world. Int. J. Forecast. **25**(4), 794–812 (2009)
30. Makridakis, S., Wheelwright, S.C., Hyndman, R.J.: Forecasting Methods and Applications. Wiley (2008)
31. Makridakis, S., Spiliotis, E., Assimakopoulos, V.: The M4 competition: 100,000 time series and 61 forecasting methods. Int. J. Forecast. **36**(1), 54–74 (2020)
32. Mann, A.: The power of prediction markets. Nat. News **538**(7625), 308 (2016)
33. McBride, M.F., Fidler, F., Burgman, M.A.: Evaluating the accuracy and calibration of expert predictions under uncertainty: predicting the outcomes of ecological research. Divers. Distrib. **18**(8), 782–794 (2012)
34. Ni, H., Chen, A., Chen, N.: Some extensions on risk matrix approach. Saf. Sci. **48**(10), 1269–1278 (2010)
35. Nikolopoulos, K., Alghassab, W.S., Litsiou, K., Sapountzis, S.: Long-term economic forecasting with structured analogies and interaction groups (No. 19018) (2019)
36. O'Donoghue, T., Somerville, J.: Modeling risk aversion in economics. J. Econ. Perspect. **32**(2), 91–114 (2018)
37. Orrell, D., McSharry, P.: System economics: overcoming the pitfalls of forecasting models via a multidisciplinary approach. Int. J. Forecast. **25**(4), 734–743 (2009)
38. Pasman, H.J., Rogers, W.J.: How to treat expert judgment? With certainty it contains uncertainty! J. Loss Prevent. Process Ind. 104200 (2020)
39. Paté-Cornell, E.: On "black swans" and "perfect storms": Risk analysis and management when statistics are not enough. Risk Anal.: Int. J. **32**(11), 1823-1833 (2012)
40. Phadnis, S.S.: Effectiveness of Delphi-and scenario planning-like processes in enabling organizational adaptation: a simulation-based comparison. Futures Foresight Sci. **1**(2), e9 (2019)
41. Redelmeier, D.A., Ng, K.: Approach to making the availability heuristic less available. BMJ Qual. Saf. **29**, 7 (2020). https://doi.org/10.1136/bmjqs-2019-010079
42. Restocchi, V., McGroarty, F., Gerding, E.: The stylized facts of prediction markets: analysis of price changes. Physica A **515**, 159–170 (2019)
43. Rezaie, K., Amalnik, M.S., Gereie, A., Ostadi, B., Shakhseniaee, M.: Using extended Monte Carlo simulation method for the improvement of risk management: consideration of relationships between uncertainties. Appl. Math. Comput. **190**(2), 1492–1501 (2007)
44. Rowe, G., Wright, G.: The impact of task characteristics on the performance of structured group forecasting techniques. Int. J. Forecast. **12**(1), 73-89 (1996)
45. Surowiecki, J.: The Wisdom of Crowds. Anchor (2005)
46. Taleb, N.N.: The Black Swan: The Impact of the Highly Improbable, vol. 2. Random house, US (2007)
47 Tetlock Philip, E.: Expert Political Judgment: How Good Is It? How Can We Know. Princeton: Princeton University Press (2005)
48. Unal, R., Keating, C.B., Chytka, T.M., Conway, B.A.: Calibration of expert judgments applied to uncertainty assessment. Eng. Manag. J. **17**(2), 34–43 (2005)
49. van Dijk, D., Franses, P.H.: Combining expert-adjusted forecasts. J. Forecast. **38**(5), 415–421 (2019)
50. Ripley, B., Venables, B., Bates, D.M., Hornik, K., Gebhardt, A., Firth, D., Ripley, M.B.: Package 'mass'. CRAN (2013)
51. Velasquez, E.D.R., Albitres, C.M.C., Kreinovich, V.: Measurement-type "Calibration" of expert estimates improves their accuracy and their usability: pavement engineering case study. In: 2018 IEEE Symposium Series on Computational Intelligence (SSCI), pp. 301–304. IEEE (2018)

52. Walker, K.D., Catalano, P., Hammitt, J.K., Evans, J.S.: Use of expert judgment in exposure assessment: Part 2. Calibration of expert judgments about personal exposures to benzene. J. Exposure Sci. Environ. Epidemiol. **13**(1), 1–16 (2003)
53. Wolfers, J., Zitzewitz, E.: Five open questions about prediction markets. NBER Working Paper No. w12060. National Bureau of Economic Research, US (2006)
54. Wright, G., Goodwin, P.: Decision making and planning under low levels of predictability: enhancing the scenario method. Int. J. Forecast. **25**(4), 813–825 (2009)
55. Zhang, Y., Cao, K., Yang, Z., Liang, K., Cai, Z.: Risk and economic evaluation of aircraft program based on Monte Carlo simulation. J. Aircraft 1–9 (2021)

Chapter 6
Humans in the Network

6.1 Introduction

In the early 1950s, a hugely influential and combative French economist and engineer, Maurice Allais, became so agitated with how the economics mainstream treated human behavior that he decided to fly a challenge at the foundations of the then dominant utility theory. The challenge came in a 1953 publication provocatively titled *The Foundations of a Positive Theory of Choice involving Risk and a Criticism of the Postulates and Axioms of the American School.*[1] The misplaced postulates and axioms of what he called the "American school" were the von Neumann-Morgenstern axioms that defined the very heart of utility and imposed strict assumptions on individual preferences. One of these was the independence axiom—the idea that people will be indifferent between two choices and those same two choices with an added constant component. Allais crafted an apt experiment showing that this clearly fails in practice and thus showed that the independence axiom is not tenable. In his own tongue-in-cheek words, the French economist was concerned not about the "rational" but about the "real" human.[2]

His insight had a profound impact and as a result nothing changed for the next two decades. The economics profession largely continued using the von Neumann-Morgenstern axioms and significantly expanded the use of utility theory and its workhorse, the utility function, in economic applications of increasing complexity. It was it the 1970s when the Allais paradox and similar insights were rediscovered as work begain in earnest on a new strand of inquiry—behavioral economics. Pioneered by psychologists Daniel Kahneman and Amos Tversky, lab experimentation on human choice provided more and more empirical insight on how "real" people make decisions, how they conceptualize and deal with risk, and how this can be formally modeled for the needs of economists, business researchers, and practitioners. Apart from the expected insight that people are almost never strictly rational,

---

[1] Allais [2].

[2] Mongin [39].

A. Gerunov, *Risk Analysis for the Digital Age*, Studies in Systems, Decision and Control 219, https://doi.org/10.1007/978-3-031-18100-9_6

work on behavioral economics demolished the myth of monolithic risk preferences such as risk aversion across all decision situations and domains. The unitary idea behind risk preferences is further undermined in that there are significant differences in risk perceptions not merely among individuals but also between different countries at different levels of economic development.[3]

In fact, we now have the understanding that people are, in the famous words of psychologist Dan Ariely, predictably irrational. They tend to misconstrue objective probabilities and instead form subjective evaluations of the risks they face. These subjective probabilities, in conjunction with other decision biases and the structure of the decision situation (i.e., the decision architecture) critically impact risk perceptions and significantly modify human decisions. Putting the decision-maker in a digital environment adds an additional layer of complexity. Throughout their online activity humans are continuously tracked and most of their actions are logged. This data is used to create specific decision architectures, known in the parlance of marketing as user experience, which skews human perception and guides deliberation. Positioning in the online social networks further sharpens peer comparisons and may give rise to herd behaviors and groupthink. Finally, the issue of information overload begins to be more and more prominent when analyzing choice and risk perceptions.

The quest for formally modeling risk preference has come a long way from crudely formulated utility functions to more sophisticated versions that are based on extensive research. The perfect way to model preferences has not yet been reached and each of the many proposals has some shortcomings. However, the practical imperative to understand how human beings perceive risk and act under uncertainty is probably stronger than ever, necessitating the use of even imperfect tools to that end. This chapter aims to present the state of the art in this respect and also some novel experimental results on human behavior in an experimental online environment.

The next section begins with an overview of classical modeling of risk preference as embodied in the numerous utility functions prevalent in economic theory. The following topic is the subjective perception of risk and its formalization that draws upon insights from Prospect Theory. The chapter proceeds to review newer insights on biases and the construction of decision architectures that modulate risk perceptions and guide decision-making. Online human behavior is reviewed next, showing how online environments, and particularly social networks, amplify some biases and lead to suboptimal decisions and decreases in utility. This is then followed by presenting data on an artificial online social network (OSN) showing how human subjects' decisions under risk are affected by changing the level of information exposure and providing mechanisms for peer influence. Those topics are relatively new and research is yet to reach a consensus on many issues but first indicative results are both enlightening and actionable.

[3] l'Haridon and Vieider [31].

6.2 Utility and Risk Preferences

The utility function has been the traditional go-to tool when modeling individual preference.[4] It aims to formalize the link between given input parameters (e.g. consumption) to the satisfaction derived from them (i.e., the resulting utility). In its implicit form it shows utility u as a function of input parameters x_i of the following form:

$$\mathrm{u} = u(x_i) \tag{6.1}$$

The standard orthodoxy assumes that satisfaction is non-decreasing in inputs, which yields a positive first derivative. On the other hand, it assumes that there is decreasing satisfaction in additional increases of input parameters, leading to a negative second derivative. Should this be the case, this implies a relative risk aversion on the part of the individual. However, the general implicit function is of limited use for practical applications, as explicit forms are needed to generate quantitative predictions and to study risk attitudes in a rigorous way. Many proposals for such forms have proliferated over the past seventy years and few of them are in wide use. One of the dimensions along which these by be distinguished is the type of risk aversion displayed. Absolute risk aversion (ARA) measures how an agent's risk preferences change in response to changes in wealth, w. ARA is thus defined by the following:

$$ARA(w) \equiv A(w) = -\frac{\frac{d^2u(w)}{dw^2}}{\frac{du(w)}{dw}} = -\frac{u''}{u'} < 0 \tag{6.2}$$

Should an agent be willing to increase his or her risk exposure (e.g. by investing more in risky assets) with the increase of their wealth, then they are characterized by a decreasing ARA. This is supposed to be more rational than the alternatives of constant or increasing ARAs. On the other hand, if the agent is not looking at the absolute level of wealth but rather at incremental relative changes, then one considers relative risk aversion (RRA). The definition for RRA is as follows:

$$RRA(w) = (w)[ARA(w)] \tag{6.3}$$

Thus, if the decision-maker who sees an increase in wealth is increasing the risk exposure (e.g. by investing in risky assets) by a greater proportion of the wealth, then RRA is decreasing. Under this specification, agents are more risk-seeking as they have greater wealth. Decreasing RRA is considered more realistic and is more widely used than either of the other two RRA versions or any ARA specification. The next sections proceed to show explicit utility functions embodying either the ARA or the CRA assumption.

---

[4] Harel et al. [26].

6.2.1 Constant Absolute Risk Aversion (CARA)

A first pass in formally modeling risk preference is to understand constant risk aversion. Typically, functions of the CARA family utilize exponential utility forms, whereby a certain parameter governs the level of risk preference.[5] The explicit form for the exponential utility $u(x)$ for the simplest case of one input parameter x and a degree of risk aversion γ is the following:

$$\mathrm{u}(x) = \begin{cases} 1 - e^{-\gamma x}, & \gamma \neq 0 \\ x, & \gamma = 0 \end{cases} \tag{6.4}$$

The measure for risk preference or the level of absolute risk aversion here does not depend on the amount of initial wealth—in fact, it is a constant. It is equal to:

$$ARA = -\frac{u''}{u'} = \gamma \tag{6.5}$$

This means that at any levels of wealth or initial endowment, the individual will only apportion a constant amount of it towards a risky choice. In the typical example of an investment decision between a risky and a riskless asset, individuals with CARA utility will always set a certain absolute amount of resources towards the risky asset. By extension, when their wealth increases they will invest any additional increase towards the riskless asset, and none—towards the risky alternative. This has led many economists to consider the implications of CARA utility to be unrealistic as there is ample empirical data that investors tend to undertake more risky positions as their wealth or endowment grows. Intuitively, agents who can afford to take up more risk actually do so.

Still, a CARA utility function is able to generate different types of risk preference as the γ parameter is changed. If the value of this exponent is greater than one, then risk-averse behavior is observed and individuals derive less utility from taking on more risk. Should the exponent γ be less than one, this describes risk-seeking behavior. For values of $\gamma = 0$, the behavior elicited is the risk-neural one. Varying this parameters, the analyst obtains different curvature of the utility functions and is thus able to describe different types of behavior (Fig. 6.1).

While an initially meaningful and important step in modeling preferences, CARA utility functions are nowadays considered less useful and thus few sophisticated applications take them in earnest.[6] They may still be taken as a benchmark case or added as a comparison as they are mathematically tractable and well-known but their key failure to model increased risk-taking with increased wealth is an unsurmountable deficiency. This is also visible in problems of empirical fit to data that yield hugely varying estimates of the ARA, and outright eliminating CARA specifications.[7] Thus,

[5] Meyer [38].

[6] Harel et al. [26].

[7] Babcock [7], Eisenhauer and Ventura [19], Guiso and Paiella [25].

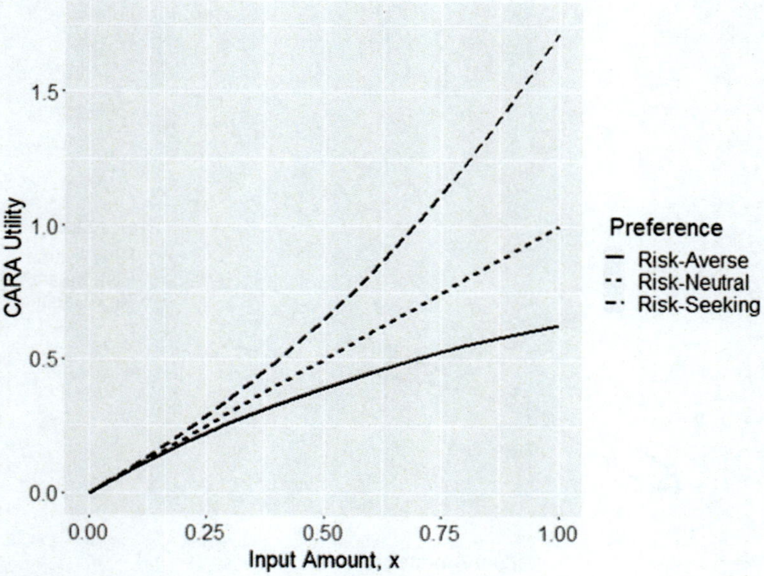

Fig. 6.1 CARA utility functions, exhibiting different risk preferences, with $\gamma = -1$ for risk seeking behavior and $\gamma = 1$ for risk aversion behavior

the analyst is advised to be aware of CARA but refrain from using it unless the situation supposes an invariant wealth level. Some observers have proposed firm decisions as a possible application but arguments from economic theory show that a firm that is not credit-constrained need not maintain a constant level of capital. The analyst is thus advised to resort to more sophisticated specifications of the utility function.

6.2.2 Constant Relative Risk Aversion (CRRA)

As an improvement upon the criticisms to the CARA family of utility functions, one may instead use a specification in which the degree of risk aversion increases with wealth but the relative amount of risk aversion remains at a constant. This usefully captures some empirical behavior as agents with greater means tend to have greater risk exposures. At the same time, however, the proportion of endowment that is spent on risk remains the same. This is well captured by the CRRA family and a typical representative is the isoelastic utility which is of the following form[8]:

$$\mathrm{u}(x) = \begin{cases} \frac{1}{1-\gamma} x^{1-\gamma}, \ \gamma \neq 0 \\ ln(x), \quad \gamma = 0 \end{cases} \tag{6.6}$$

[8] Meyer [38].

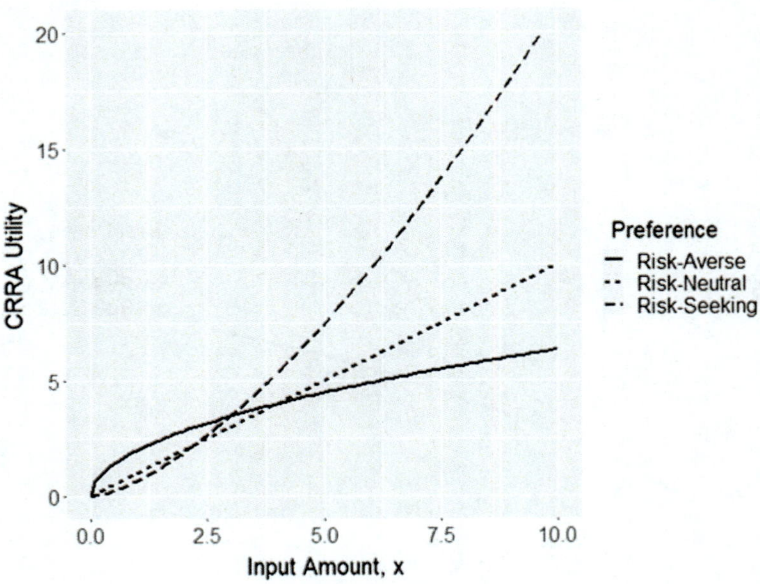

Fig. 6.2 CRRA utility functions, exhibiting different risk preferences, with $\gamma = -0.5$ for risk seeking behavior and $\gamma = 0.5$ for risk aversion behavior

Here γ gives the degree of relative risk aversion and thus the risk aversion measure $A(x)$ is given by the following:

$$\mathrm{A}(x) = \frac{\gamma}{x} \tag{6.7}$$

While the CARA specification allows the analyst to precisely specify the magnitude of risk aversion of the given agent (e.g., showing how much risk an investor is willing to take), there is no possibility to set the slope of the risk aversion measures. Thus, one cannot model change. In the CRRA instance, the slope of the relative risk aversion is zero, while the slope (and thus the decrease) of the absolute risk aversion is $-\gamma/x^2$. Still, this utility function is able to capture different types of behaviors that may be of use for practical applications[9] (Fig. 6.2).

Such specifications of the utility function are particularly prevalent in investment and finance research as they pose a sufficiently good tradeoff between costs and benefits. On the one hand, CRRA utility are mathematically tractable and relatively simple to calculate, with a clearly defined and straightforward coefficient of risk aversion driven by a single parameter, γ. Analysts are generally well aware of them and they are tested and applied in numerous applications in both microeconomic and macroeconomic problems.[10] CRRA utility functions also capture some stylized

[9] Wakker [61].

[10] Ñíguez et al. [42].

facts of behavior. In some modeling situations those advantages outweigh the relative lack of realism and rigidity of imposing one single simple risk preference on agents independent of the decision situation.

6.2.3 Hyperbolic Absolute Risk Aversion (HARA)

More sophisticated specifications of the utility functions try to address the limitations and constraints evident in simple CARA and CRRA formulations. One of the most popular classes in use is the family of Hyperbolic Relative Risk Aversion functions that allow one to model the dynamics changes of absolute and relative risk aversion. The HARA utility function is of the following form:

$$\mathrm{u}(x) = \frac{(1-\gamma)}{\gamma}\left(\frac{\beta x}{1-\gamma} + \eta\right)^{\gamma} \tag{6.8}$$

This function is characterized by three parameters that with γ driving the risk preferences and β and η being tuning parameters. The parameters themselves are subject to the following restrictions:

$$\begin{aligned} &\frac{\beta x}{1-\gamma} + \eta > 0 \\ &\gamma \neq 1 \\ &\beta > 0 \end{aligned} \tag{6.9}$$

The initial conception of the HARA utility, as with many other utility functions, was through assumption. Later, it was argued that it is not only a convenient way to approach decision problems but stems directly from the logic of optimization in microeconomic problems.[11] At usually, one can study the risk preferences inherent in this specification by taking recourse to the absolute risk aversion measure $A(x)$. In the case of HARA, it is the following:

$$\mathrm{A}(x) = \frac{1}{\frac{x}{1-\gamma} + \frac{\eta}{\beta}} \tag{6.10}$$

It is quite clear from Eq. (6.10) that by changing the respective parameter values the analyst may obtain not only different risk preferences but also different slopes of the absolute and relative risk aversion. This versatility allows the modeling of a wide range of more realistic human behaviors as risk preference change in the vicinity of certain values. Insights from behavioral economics show that this is indeed the case as people may have different preference for small- and for large-stakes bets. Naturally, this flexibility and power comes at the expense of tripling the parameters

[11] Perets and Yashiv [45].

that govern the function and the analyst needs to make a careful evaluation of whether the increase in complexity justifies a possible increase in explanatory power.

A particular attraction to using HARA is that simpler CARA and CRRA utility specifications may be obtained by setting the parameters in Eq. (6.10) to specific values. If the parameter γ goes to infinity, then the CARA specification from Eq. (6.4) results. If, on the other hand, $\eta = 0$, then the CRRA form is obtained as in Eq. (6.6). In a similar vein to CRRA finite values of $\gamma < 1$ imply that decision-makers will increase their exposure to risk as their initial endowment or wealth increases. Thus, an investor with HARA utility and γ of less than unity will invest more in risky assets with the increase in disposable funds. Empirically, this is considered the baseline case, and thus defines one of the key parameterizations of the HARA utility function.

HARA is also preferable on the grounds that it generates concavity of the consumption functions that is useful for economic modeling.[12] While the quest for a perfect fit to empirical data remains elusive for virtually all utility function alternatives, it seems that the full HARA specification may provide a sufficient level of empirical fit even across very heterogenous decision-makers.[13] This is probably due to the fact that it captures two key behavioral traits in real-life: the simultaneous operation of decreasing absolute risk aversion but an increasing relative risk aversion.

6.2.4 Advanced Utility Specifications

Over the past decades using a utility function to specify individual preferences for a diverse set of economic and business problems turned into a standard approach in research and practice. This has led to the expected consequence of utility functions being tested against real-life empirical data and failing on at least some account. In turn this has given rise to a proliferation of new more or less complex specifications of the utility function, most of which are assumed a priori and only after that tested rigorously. Naturally, an overwhelming proportion of the proposed novel formulations fell through the cracks but a few fared better. Among the relatively more successful ones one should note the family of expo-power (EP) utility functions, the power risk-aversion (PRA), and the flexible three parameter specification (FTP).

In 1993 Saha introduced the expo-power utility function, hailing it as a particularly flexible form of the utility function[14]. It has no less than three tuning parameters and is of the following form:

$$\mathrm{u}(x) = \theta - exp(-\beta x^{\gamma}) = \theta - \frac{1}{e^{\beta x^{\gamma}}} \tag{6.11}$$

Some parameter restrictions are imposed here as follows:

[12] Toda [57].

[13] Brocas et al. [11].

[14] Saha [47].

$$\begin{aligned} \theta &> 1 \\ \beta\gamma &> 0 \end{aligned} \tag{6.12}$$

One needs to keep in mind that θ is an additive constant so its rather unimportant in maximization problems as it become zero in the derivative, which makes the specification in Eq. (6.11) essentially a two-parameter function for all practical purposes. The analyst may obtain the desired type of risk preference (aversion, seeking, neutrality) as well as the desired slope of the risk aversion measures by setting appropriate values to the parameters β and γ. For example, the analyst may obtain constant absolute risk aversion by setting $\gamma = 1$ and $\beta > 0$, increasing absolute risk aversion with $\gamma > 1$ and $\beta > 0$, and decreasing absolute risk aversion with $0 < \gamma < 1$ and $\beta > 0$.[15]

The absolute risk aversion measure of this utility function is relatively straightforward and is of the following form:

$$\mathrm{A}(x) = \frac{1 - \gamma + \gamma\beta x^{\gamma}}{x} \tag{6.13}$$

While the expo-power utility is somewhat more complex that CRRA and CARA forms, it has the added benefit that it can account for a wide range of risk attitudes and preferences that can be obtained by inserting different parameter values. In this respect it is similar to HARA. One more benefit is that it contains familiar specifications, thus reducing to CARA from with finite parameter values. At the same time, it remains tractable and useful, still featuring a compact risk aversion measure. The great drawback is that is has been subjected to rather limited empirical testing and thus its realism remains suspect.

Independently of Saha, in 2000 Xie[16] proposed another similar specification of the utility function that he called the power risk-aversion (PRA) one. It is also supposed to provide significant flexibility to the modeler in capturing different types of risk preferences and their respective dynamics (and thus the curvature of the utility function). The PRA utility is of the following from:

$$\mathrm{u}(x) = \frac{1}{\alpha}\left\{1 - exp\left[-\alpha\left(\frac{x^{1-\sigma} - 1}{1 - \sigma}\right)\right]\right\} \tag{6.14}$$

The two parameters—α and σ—are both restricted to take non-negative values. The measures of risk aversion can be set to increase of decrease by choosing appropriate values for the α and σ parameters. It is argued that PRA utility is also an improvement over HARA as the function remains well-defined for a wider range of inputs than it is for HARA. In addition to its flexibility and behavior, the specification in Eq. (6.14) also yields a (still) relatively simple risk aversion metric:

---

[15] Settlage and Preckel [49].

[16] Xie [65].

$$\mathrm{A}(x) = \frac{\sigma}{x} + \alpha x^{-\sigma} \tag{6.15}$$

On the theoretical front, PRA utility has, in fact, the same functional form as the EP specification. When one sets the parameters in PRA as follows: $\sigma = 1 - \gamma$, and $\alpha = \gamma\beta$, then the functional form in Eq. (6.11) is obtained. On the empirical front, PRA utility is still of limited use with it being more prominent as "one of many" specifications to be tested, rather than as the single representation of preferences.[17]

Yet another specification of a utility function aims to further generalize the EP/PRA form and thus incorporate most special cases into a single formulation with a number of versatile parameters. This was proposed by Conniffe[18] under the name flexible three parameter (FTP) utility function and is described by the following expression (with γ, k, σ being the tuning parameters):

$$\mathrm{u}(x) = \frac{1}{\gamma}\left\{1 - \left[1 - k\gamma\left(\frac{x^{1-\sigma} - 1}{1-\sigma}\right)\right]^{\frac{1}{k}}\right\} \tag{6.16}$$

In order to avoid imaginary numbers for non-integer values of $1/k$, the following restriction must be imposed:

$$1 - k\gamma\left(\frac{x^{1-\sigma} - 1}{1-\sigma}\right) > 0 \tag{6.17}$$

The key advantage in using the FTP specification is that if needed parameter values can be chosen in such as way that the function is bounded from below or from above. In the former case, the lower bound describes phenomena such as the subsistence level of income or wealth. Similarly, the latter case reflects phenomena such as saturation with consumption or income. Furthermore, the FTP utility shares some important characteristics with HARA utility such as the ability to model different behaviors and their dynamics. Finally, FTP can be considered a general case of other proposed popular utility specifications. For example, setting the parameter k to zero obtains the PRA utility function in Eq. (6.14). The risk aversion of the FTP utility is somewhat more complex than the ones already reviewed. In has the following form:

$$\mathrm{A}(x) = \frac{\sigma}{x} + \frac{(1-k)\gamma x^{-\sigma}}{1 - k\gamma\left(\frac{x^{1-\sigma}-1}{1-\sigma}\right)} \tag{6.18}$$

The complexity of the FTP specification is its major drawback. On the one hand, the need to specify three parameters is challenging. Fitting them from empirical data is not always straightforward and testing their subsequent empirical validity is done through testing of joint restrictions. It is thus unclear to the analyst what the

---

[17] See e.g. Dowd et al. [18].

[18] Conniffe [16].

key drivers of the results are, and whether the model itself is valid. Moreover, the complex form of both the function and its risk aversion measure make it less tractable and less convenient to work with. This is a significant issue particularly against the backdrop of the lack of empirical justification of this form.

While FTP is able to cover a significant number of plausible behaviors, it remains unclear what the actual behaviors are, and what parameterization will include a general representative case. One can make the argument for a consensual set of values that collapse the specification in Eq. (6.16) to simpler functions such as PRA or even versions of CRRA. If that is the course of action chosen, it still remains unclear why the analyst should start from the more sophisticated FTP formulation only to water it down for practical purposes. All those issues have marred the adoption of the FTP form and it is only rarely used in research and practical applications, with most modelers preferring simpler specifications such as the CRRA or HARA utility functions.

6.2.5 *The Utility of Utility Functions*

The utility function is essentially a strange animal. It purports to capture human behavior and encode risk preferences but almost all formulations are divorced from it—i.e., utility functions are not derived from observed empirics but most often are just assumed by decision scientists.[19] This substitution of actual behavior for assumed plausible one poses numerous important tradeoffs when using those formulations for actual risk modeling and evaluation. First, there is no single optimal formulation of the utility function of a given individual, organization, or other entity such as a state. In fact, there are a number of popular and plausible interpretations that are currently in use, and those have different properties especially when it comes to modeling risk preferences. Popular utility functions are listed in Table 6.1 and while it seems that the CRRA specifications are more prevalent, all of the enumerated ones (and more) are in use. It is difficult to determine which one type of utility is best for a modeling situation. While it is possible to fit all of them against empirical data on risky choice, research has shown mixed results on their performance. What is more, more sophisticated utility function formulations such as the power risk-aversion (PRA) one necessitates the simultaneous fit of numerous parameters and joint restriction testing which makes it all the more difficult to compare to other specifications and outline a clear winner.

Second, often the choice of a utility function depends not on how realistic it is, but on how tractable it turns out to be for deriving a numeric solution for a specific decision problem. It is thus hardly surprising that the CRRA specifications are a preferred tool of choice—their mathematical elegance, simple optimization, and straightforward measure of risk aversion make them easy to plug into a complex problem mathematically described by a number of equations that can be eventually solved.

[19] Meyer [38].

Table 6.1 Common utility functions and their risk aversion measures

#	Utility function	Typical form	Type of risk aversion	Risk aversion measure, A(x)
1	Constant absolute risk aversion, CARA	$1 - e^{-\gamma x}$	Constant in wealth	γ
2	Logarithmic utility	$\ln x$	Constant in wealth, special case of CARA	$\frac{1}{x}$
3	Constant relative risk aversion, CRRA	$\frac{x^{1-\gamma}}{1-\gamma}$	Relative to wealth	$\frac{\gamma}{x}$
4	Hyperbolic absolute risk aversion	$\frac{(1-\gamma)}{\gamma}\left(\frac{\beta x}{1-\gamma} + \eta\right)^{\gamma}$	Can be both relative or constant to wealth with changing behavior (slope) of $A(x)$	$\frac{1}{\frac{x}{1-\gamma} + \frac{\eta}{\beta}}$
5	Quadratic utility	$x - \beta x^2$	Increasing absolute risk aversion, special case of HARA	$\frac{2\beta}{1-2\beta x}$
6	Expo-power utility	$-\exp(-\beta x^{\gamma})$	Can be both relative or constant to wealth with changing behavior (slope) of $A(x)$	$\frac{1-\gamma+\gamma\beta x^{\gamma}}{x}$

Ease of application has sometimes been a major driver behind the choice of a specific formulation and while this may seem questionable from a methodological standpoint, it is often imperative from a practical one. An additional concern when leveraging overly complicated formulations is that they call for simultaneously assuming or fitting numerous parameters. Since it is often possible to find a number of sets of values that satisfy the data or the imposed constraints, it remains unclear which set is the true one, leaving the problem poorly defined. All those issues lead the analyst to be strongly in favor of simple and tractable formulations of utility (if at all needed).

Third, the proposed utility functions suppose a single type of behavior across domains and, sometimes, across outcomes. Essentially, this means that irrespective of the choice situation (e.g. financial markets investment, medical decisions, simple gambles) and its parameters (outcomes, tradeoffs and probabilities), individuals will tend to exhibit a similar set of risk preferences. On the one hand, this makes decision-making consistent and much easier to model. On the other hand, results in psychology and behavioral economics clearly show that individuals are sensitive towards context, problem formulation, type of outcome and magnitude of probabilities.[20] This means that a given utility specification may perform well under a set of specific assumptions in a given situation but underperform as change occurs. The analyst thus needs to factor in that optimal utility specifications may not be portable across decision problems. As a partial solution to this some researchers have tended to check their results leveraging a number of popular utility functions, thus checking the sensitivity of

[20] For a sweeping overview, please refer to Kahneman [27].

obtained insights to changes in the specifications. This works well when conclusions remain the same, but this is hardly guaranteed.

Fourth, and despite their many shortcomings, utility functions are still able to capture major stylized facts of observed economic behavior. They reflect the tendency of individuals to avoid risk, and more sophisticated formulations even allow the change of the rate of risk avoidance as the magnitude of the expected outcome changes. The risk aversion measures of those utility specifications allow the formal comparisons of judgments, courses of action or choices in terms of their perceived level of risk. This allows a utility function to be used for practical problems such as portfolio selection *a la* Modern Portfolio Theory, labor supply, insurance, and pension funding. The familiarity of the expert community and the fact that the utility functions' properties are well understood ease not only their application but also the communication of the results derived. While imperfect at the individual level, utility function specifications may be able to provide a good first approximation to risk preference at the aggregate level.

In conclusion, simple utility functions can be shown to suffer from multiple shortcomings, mainly stemming from the need of simple formulations and their divorce from the complex reality of actual decision-making. This makes them practically useless to understand decisions at the individual level where the layers of psychological complexity take precedence. However, utility functions may still be useful to model behavior in a wide set of decision settings such as aggregate behavior, estimating benchmarks, and conducting simulations. Mathematical specifications of preferences may be unsuitable for individuals but at the group level much randomness is cancelled out. This means that a well-formulated utility function can be used to approximate the *aggregation of the results* stemming from disparate individual preferences. This is naturally useful for modeling social and economic systems, and prescribing courses of policy action. In a similar vein, risk analysis may proceed through simulations of complex environments. Those simulations need to be equipped with a risk preference and choice mechanism that is consistent, well-understood, and tractable. Leveraging a familiar utility specification fits well with those requirements and eases further analysis. Finally, the utility function may be conceived of as a normative benchmark—the way a rational decision-maker should make choices. Having this benchmark the analyst may study deviations from it and prescribe courses of action. All in all, despite their many flaws utility functions retain certain utility for modelers. Seeing this and attempting to further enhance them by making them more realistic, psychologists and behavioral economists came to the rescue in the 1970s. Prospect Theory emerged.

6.3 Prospect Theory

Standard economic modeling of decisions under risk has traditionally revolved around the tenets and ideas of expected utility. This presupposes that decision-makers have a well-defined and tractable utility function such as the ones presented in the previous section, and that they make their judgment based on a probabilistically weighted expected utility, $E[U(x_i)]$. As outlined in Chap. 1, this is of the following form (with $f(x)$ being the probability distribution of events):

$$\mathrm{E}[U(x)] = \int_a^b u(x) f(x) dx \tag{6.19}$$

Plain vanilla expected utility postulates that decision-makers' risk preference is well captured in the utility function and then objective probabilities are used to weight each outcome, thus arriving at a determination of terminal utility. The mathematical elegance and the compelling logic and simplicity of those ideas, however, came crumbling down as they faced the reality of human choice in the lab. While questions of the applicability of this simple model have been raised before, it was only with the work by Kahneman and Tversky that behavioral economics came into its own. Those two psychologists found compelling evidence that human decisions are not rational and are instead subject to a number of heuristics and biases that skew them. In an attempt to reconcile experimental results with mainstream modeling at the time, Kahneman and Tversky proposed a reformulation of the two building blocks of standard expected utility—to substitute the utility function for a more sophisticated value function, and to move from objective to subjective probability.[21] Prospect Theory was born. It was later improved and expanded into Cumulative Prospect Theory[22] which is a natural starting point for understanding perceptions of risk and the decisions stemming from that.

The main goal behind the creation of Prospect Theory was to account for four major empirical regularities that were observed in experimental settings.[23] First, subjects displayed a significant degree of reference dependence, i.e., people cared where a given risky situation would land them relative to their initial levels of wealth or endowment. Essentially, they derive utility from gains and losses relative to some reference point they have in mind. Beside the experimental results, this reflects psychological characteristics of human beings who consider changes (in temperature, light, wealth, etc.) as more salient than the absolute levels of those same variables.

Second, loss aversion was a major factor in individual behavior, and it was even more pronounced than traditional utility models at the time could adequately explain. In essence, people will lose more utility from losing a certain amount x, than they will gain utility from attaining the same amount. Third, there was a diminishing sensitivity to risk. This reflects the finding that people tend to be risk averse in the case of medium stakes, but they change their behavior abruptly when it comes to losses—they suddenly become risk-seeking. This would call for specific curvature of the utility function—concavity in the region of gains, and convexity in the region of losses.

---

[21] Kahneman and Tversky [28].

[22] Tversky and Kahneman [58].

[23] Barberis [8].

Fourth, the last pillar of Prospect Theory is probability weighting. It seeks to explain the fact that people do not seem to use objective probabilities in their decisions, but they skew them, which has implications on how they perceive risk and uncertainty. Thus, a fuller explanation of the behavior of "real" people necessitates both a reformulation of the utility function, and a mechanism for turning objective probability into subjective probability. Kahneman and Tversky propose both—these are the value function and the probability weighting function. Those simple formulations have revolutionized the way risk is modeled and decisions are conceptualized. The value function *v(x)* proposed is of the following form:

$$v(x) = \begin{cases} x^{\alpha}, & x \geq 0 \\ -\lambda(-x^{\alpha}), & x < 0 \end{cases} \tag{6.20}$$

A key difference in the value function is that it has different curvature for the case of gains versus losses. This reflects the behavioral trait that people feel differently when it comes to winning as opposed to losing. An evolutionary explanation is usually advanced to account for this phenomenon—in primordial times our ancestors needed to be much more attuned to potential negative risks (e.g. a giant predator in the ambush) than to positive ones (e.g. edible fruit). The former meant the difference between life and death, while the latter was just nice to have. By extension, the modern human finds losses much more salient than gains. This is reflected in the formulation expressed in Eq. (6.20) that has an inflection point at $x = 0$, as the individual goes from one domain to the other. Using experimental data, Kahneman and Tversky[24] are also able to estimate the parameters of this new utility (value) function to be $\alpha = 0.88$ and $\lambda = 2.25$, thus making their specification complete. This function is presented graphically in Fig. 6.3.

The introduction of a new utility function is insufficient to fully account for realistic individual behavior. A key observation from the lab is that while people should use objective probabilities for decision-making, they tend to pass them through the lens of subjective evaluation, thus skewing them. Sometimes those differences may be quantitatively significant and of marked practical importance. Prospect Theory thus proposes a weighting equation that allows for the conversion of objective into subjective probabilities. It is of the following form:

$$\pi(p) = \frac{p^{\delta}}{\left(p^{\delta} + (1-p)^{\delta}\right)^{\frac{1}{\delta}}} \tag{6.21}$$

The formulation in Eq. (6.21) still accounts for the fact that people tend to approach the probabilities of gaining and losing in different ways. Thus, the tuning parameter δ has different values depending on the types of outcomes. For positive outcomes $\delta_+ = 0.61$, and for negative ones it is $\delta_- = 0.69$. The main difference in applying

[24] Tversky and Kahneman [58].

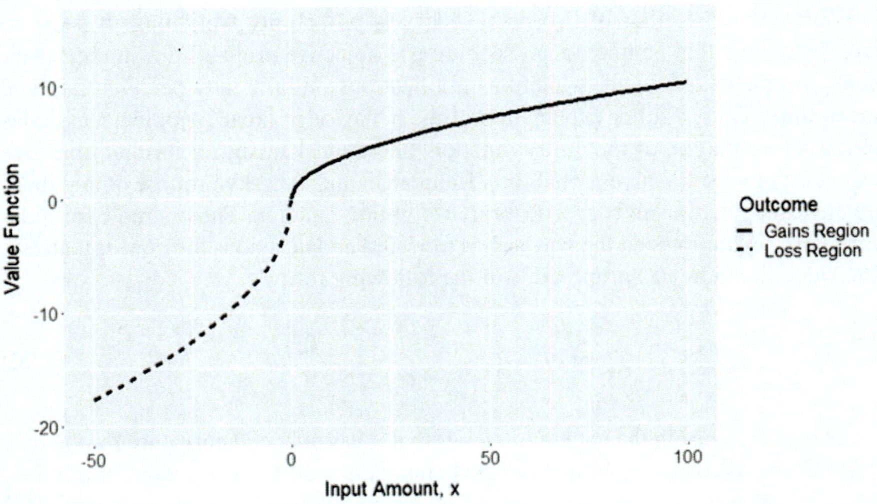

Fig. 6.3 Prospect theory value function with $\alpha = 0.5$ and $\lambda = 2.5$

weighting between the initial version of Prospect Theory and the later one—Cumulative Prospect Theory—is that in the latter weighting is applied to cumulative probability (e.g. of losing x or more, or of gaining at least x). A visual representation of a typical weighting function is shown in Fig. 6.4. It clearly reflects the tendency of humans to overweigh low-probability events but to under-weigh high-probability events, which gives rise to a nuanced and subtle set of risk preferences.

The key insight from weighting is that individuals tend to misconstrue the tails of the probability distribution. On the one hand, this makes intuitive sense as decision-makers are least exposed to rare (tail) events. On the other hand, this is of particular concern as significant and potentially catastrophic risks lurk exactly there. All in all, prospect theory has been a revelation for modeling risk attitudes.[25] Traditionally, the form of the utility function has been assumed by modelers based on anecdotal observations, plausible behavior and first principles of decision-making. Experimental work on decision-making has amply shown that those assumptions are often unrealistic, and humans tend to have a very complex approach to understanding and conceptualizing risk that yields a number of deviations from rationality. Crucially, human perception need not coincide with objective reality and key risk features such as probability of occurrence are perceived in a skewed way. Even the subjective interpretation of risk varies significantly depending on the type of risks, with people being more focused on losses than gains.

One would think that those insights would have crucial repercussions for modeling risky situations in economics, business, and society. Despite some recent advances, this has largely not been the case for decades.[26] Even simple and tractable additions

---

[25] O'Donoghue and Somerville [43].

[26] Barberis [8].

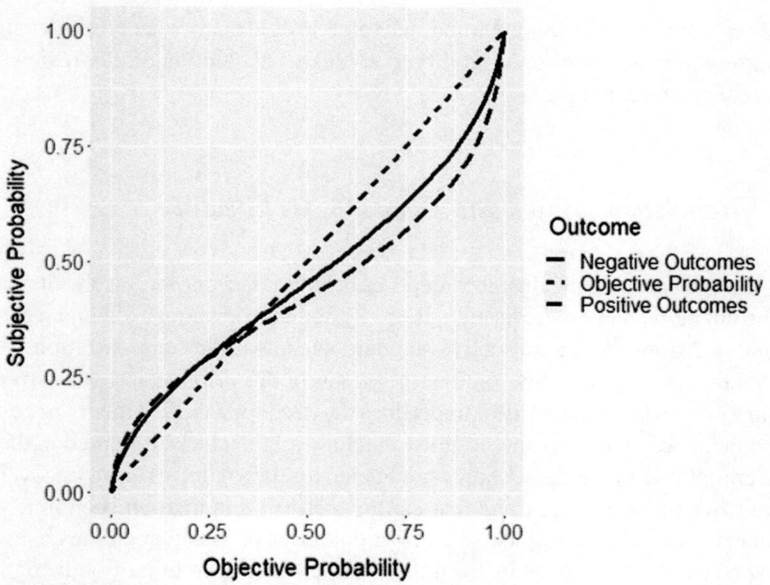

Fig. 6.4 Prospect theory weighting function with $\delta_+ = 0.61$ and $\delta_- = 0.69$

to Expected Utility Theory such as those in Prospect Theory take a long time to be whole-heartedly adopted and their application lags significantly behind. The main areas where Prospect Theory is used are financial investments and insurance, and is slowly making headway into research on consumption and savings, industrial organization, and labor supply. Macroeconomic applications are few and far between, and it seems that the most promising venue for their application is agent-based modeling, whereby a large number of heterogenous agents are simulated in order to aggregate their behaviors into resultant economic dynamics.

6.4 Neural and Behavioral Foundations

The elegant formulations of Prospect Theory do much to enable a deeper understanding of human decisions under risk. However, they still remain a highly stylized description of behavior under risk that fails to account for some empirically observable traits. In the vein of Kahneman and Tversky, research has continued to unveil new and new heuristics people use to make uncertain calls, and new and new biases they are subject to. While all are enlightening in their own right, some have crucial repercussions as to the risk perceptions. A further line of inquiry underlines that individual decision do not take place in vacuum but instead are crucially influenced by the context—i.e. the decision architecture. Given a specific decision architecture, individuals may be gently steered (nudged) towards a certain decision. Finally, deeper exploration of the human mind may give insight as to what engender such a wide variety of biases and susceptibility to context. This section briefly reviews key results in those research programs.

6.4.1 Heuristics, Biases and Decision Architecture

An oft-thrown comment against the adepts of behavioral economics is that its results are often not more than a mere list of biases—systematic deviations of human perceptions and decisions from a normative standard such as utility maximization. While this is vehemently denied, this comment serves as both a critique and as a surprising compliment. On the critique side it underlines the need for a unified theory of perception and behavior under risk that is able to include all the quirks observed in the lab. On the complimentary side, it shows how behavioral economists have uncovered a treasure trove of behavioral traits that can importantly impinge on economic decisions under risk and uncertainty. The effect on biases on risk perceptions is a well-established result[27] which stems from the simultaneous operation of both cognitive and emotional biases on the part of the decision-maker.[28]

Cognitive biases are in essence observed when decision-makers show systematic deviations between their responses and a normatively correct one in a judgment task. They hail from the great tradition of asking experimental subjects to evaluate probabilities and make probabilistic judgments, while in the background psychologists armed with the Bayesian theorem calculate the correct answer. The two numbers are compared, and a statistically significant difference often emerges. From these humble beginnings, behavioral economists and psychologists have proceeded to uncover and document a large number of relevant biases and some of them have importance repercussions for risk attitudes. Those are as follows[29]:

- *Anchoring bias*—it is observed when the estimate of a numerical value is strongly influenced by an initial non-related number (the anchor) that participants were exposed to;
- *Availability bias*—occurs whenever subjects overestimate the probability of a given event just because it is more salient or easy to recall;
- *Certainty effect*—decision-makers tend to prefer certain outcomes and heavily discount uncertain ones, even when the expected outcome is much larger;
- *Equalizing bias*—sometimes humans tend to give equal probabilities or equal weights to all possible outcomes or events, thus equalizing them;
- *Gain–loss bias*—humans may change their response depending on how a given problem is framed even though the outcomes and probabilities remain invariant in all cases;

---

[27] Saposnik et al. [48], Simon et al. [50].

[28] Montibeller and Von Winterfeldt [40].

[29] Also see Montibeller and von Winterfeldt [40] and the references therein.

- *Myopic problem representation*—this bias materializes as decision-makers adopt an incomplete representation of the problem at hand, or only focus on a limited number of aspects;
- *Omission of important variables*—humans are sometimes prone to mistakes, overlooking a pertinent variable that is relevant to the issue under consideration;
- *Overconfidence*—this occurs when participants overestimate actual performance or generate too narrow ranges of uncertainty (i.e. they are too confident of their estimates);
- *Proxy bias*—individuals tend to attribute more weight to intermediate proxy attributes than they do to the ones being represented by the proxies;
- *Range insensitivity and scaling biases*—decision-makers sometimes fail to properly adjust weights in response to change in the range of attributes or fail to properly account for scaling issues;
- *Splitting bias*—this is observed as the grouping or splitting of outcomes or objectives affects their importance (weights) or the probabilities attached to them.

The other large group of biases are the motivational ones—these are engendered by the level of desirability of a given outcome as viewed by the decision-maker. The textbook example is that of the nuclear engineer who always considers new nuclear power plants as being extremely low-risk. Motivational biases are also able to affect risk perceptions and provide a skewed estimate or decision. Some of the more relevant ones are as follows[30]:

- *Affect influence*—it materializes as the subject has an emotional predisposition regarding a specific event, course of action or outcome, and this influences the judgment;
- *Confirmation bias*—individuals are notoriously prone to clinging to their own ideas and beliefs and are sometimes prone to accept only evidence in support of those, and to discard evidence to the contrary;
- *Wishful thinking*—this occurs when people give higher probabilities to events that are desirable;
- *Undesirability of negative events*—in this case decision-makers provide erroneous estimates to unwanted or harmful events in the attempt to be overly cautious;
- *Ambiguity aversion*—options with explicit probabilities are preferred as opposed to those with unclear outcomes, which shares some similarity to the Certainty effect bias;
- *Base rate fallacy*—individuals may choose to ignore unconditional probabilities (base rates) and instead take cue from individual personalized information. This is also connected to the conjunction fallacy that makes people believe the joint occurrence of two events is more likely than the occurrence of the constituent events;
- *Conservatism*—this refers to the inability of decision-makers to sufficiently update their beliefs and probability estimates even after new information is received, and may be connected to the confirmation bias;

[30] Ibid.

- *Endowment effect (Status quo bias)*—individuals tend to overvalue what is in their possession or what they have invested or been involved with, thus factoring in sunk costs in their decisions;
- *Gambler's fallacy (Hot hand bias)*—people tend to include irrelevant past information in their current estimate of probability or outcomes (even for independent events);
- *Probability-related biases*—this groups consists of biases that shows that decision-makers often make mistakes when dealing with probabilities. These may include drawing conclusion from insufficient observations (sample size bias), inability to understand regression to the mean (nonregressive prediction bias) or errors when adding event probabilities (subaddivity or superaddivity bias).

By this point the reader is probably overwhelmed by the number and variety of errors humans make when assessing risky situations or being asked to pass judgment. The propositions of Prospect Theory are able to address some of those biases for the needs of modeling, but many remain standing. The usual advice is to use debiasing techniques when eliciting estimates from human experts by prompting them to think about potentially missing factors, by decomposing the decision task, providing additional information or counterfactuals, and use smart group elicitation and aggregation methods. A basic training in probabilities and an exercise in expert calibration is often an indispensable part of understanding and modeling risk perceptions.

Naturally, such long and exhaustive lists of irrational lapses on the part of human decision-makers are worrisome for mainstream expected utility theorists. Those insights from behavioral economics clearly show that individuals are not merely irrational, but in the words of Dan Ariely, *predictably* irrational[31] and are repeating the same avoidable mistakes over and over again. This may lead to a distorted view on uncertainty and risk that skews individual decisions, often to the detriment of the decision-maker. While this is the mainstream consensus, it is far from universal, with vocal and prominent critics advancing alternative interpretations. Enter Gerd Gigerenzer.

Gigerenzer maintains that there is a certain bias in the profession to see behavioral biases (the "bias bias") where there are none, and thus undermine the faith in individual decisions.[32] He marshals a significant amount of psychological research showing that individuals exhibit mostly calibrated intuitions about probabilities and are able to make adequate decisions given an intuitive problem framing and possibly some education in probability. The insight is thus that while some decisions remain imperfect, the overwhelming majority of decisions are "good enough" and the large number of biases cannot be clearly shown to be costly in terms of health, well-being or income. In a world of complex situations, it may actually be beneficial for individuals to exhibit a certain bias and resort to simple heuristics (rules of thumb) to approach uncertainty.[33]

[31] Ariely and Jones [5].

[32] Gigerenzer [23].

[33] Mousavi and Gigerenzer [41].

The key thing is to understand what situations represent knowable risks and can thus be approached with the traditional tools of expected utility; and what situations refer to fundamental uncertainty, and thus a heuristic-based approach may be better. Research has shown this to be quite effective when dealing with knowable situations with rapid feedback mechanisms, but more questionable for complex domains with delayed feedback that people are not accustomed to, such as financial markets.[34]

The failure of individuals to always conform to the lofty standards of perfect rationality due to the large number of biases they exhibit clearly points at the importance of the context they are immersed into, and the default rules that they often fall back to when unsure. In short, the decision architecture matters for the decision outcome. Given the vast scope of potential pitfalls such as anchoring and framing, the effect of starting points and base rates, as well the ambiguous individual preferences it is often crucial to think and design a decision architecture that supports beneficial choices. That is precisely the idea behind Thaler and Sunstein's urge to "nudge" people into making better decisions.[35]

Changing the context of the decision-making turns out to lead to significantly different choices in fields ranging from nutritional choices, through labor supply, and into savings and investment decisions.[36] The key behavioral insight is that people tend to rely overwhelmingly on the default choice preferring inaction over action. Often there are default choices, or at least influences from the environment, and it is up to the decision architect to gently steer (or nudge) the decision-maker towards optimal action. This powerful and controversial idea became known as libertarian paternalism and has been taken up in real-life policy in the USA and UK by creating special public sector units in charge of nudges.

6.4.2 Neural Circuitry of Risk

Decision theory in mainstream economics views preference formation and choice as relatively straightforward unitary processes that culminate in the optimization of a tractable mathematical function (utility or value function). As a stark contrast to this, insights from the lab provide a complex and deeply nuanced perspective on how people actually form preferences, understand risk, and take action. This clash brings to the forefront the question of what actually transpires during decision-making tasks and what are the low-level drivers behind certain judgments and behavior. The budding discipline of neuroeconomics holds the promise to unveiling this mystery by looking at how the brain functions and mapping the neural circuitry of risk preference and decision-making. The early promise of neuroeconomics was simple—use the advanced capabilities of functional Magnetic Resonance Imaging (fMRI) to inspect

---

[34] Puaschunder [46].

[35] Sunstein and Thaler [53].

[36] Thaler and Sunstein [55, 56].

brain activation and map different brain region to decision tasks.[37] Realizing that the drivers of brain activation are in fact electric impulses that transmit information between brain cells (neurons), the researcher can actually see how low-level electric stimulation gives rise to complex judgment. Two decades of subsequent research did reveal much insight but the overpromise of the discipline has yet to be delivered.[38]

A major breakthrough of studying the brain is the fact that during decisions people tend to activate not merely a single part of it, corresponding to an elusive construct such as the utility function, but rather numerous parts that form networks of simultaneously operating regions. Decision-making cannot thus be a unitary process in its essence but rather an emergent one. This, however, does not preclude it from being modeled as one single process for all practical purposes. Research in neuroscience and behavioral economics has largely reached the same conclusions regarding choice—people are equipped with two separate ways of approaching it.[39] Kahneman[40] refers to them as the System I and System II but other similar names abound in the literature. System I is a quick and dirty way to make decisions—it is fast, efficient, and intuitive. System I is based on emotion and dominated by decision heuristics that allow the snap evaluation of probability and the immediate choice of a course of action. It has grown out of the primordial brain and is very suited for survival in the wild. In the complex modern world it still serves man relatively well for low-stakes simple and learnable decisions. One should really not expend excessive cognitive resources to estimate the terminal expected utility of a cup of morning coffee day after day.

System II, on the other hand, is the heavy artillery of the human brain. It is slow, deliberate (and deliberative), resource-intensive, and highly efficient. System II is suited for sophisticated tasks and is often used when estimating probabilities, judging numerical quantities and making complex determinations. Homo economicus is exemplified by System II, ordinary people—not so much.[41] Evolutionary, the brain regions that are responsible for those slower and more complex processes arrived later and tend to be responsible for higher-order functions of the human brain. These are less emotional and evolved as the need to analytically process information became increasingly important for the survival of humans. Risk preferences and choices are thus a product of the simultaneous operation of those two systems. In some cases, System I takes precedence (the default option) and in some cases—System II (often after deliberate activation). A common pattern is for System I to make a decision, and System II to merely rationalize it afterwards. At any rate, one must keep in mind that again those are not monolithic unitary constructs, but rather emergent ones predicated on complex networks of neurons in different brain regions, and the connections between them.

---

[37] Camerer et al. [12].

[38] Konovalov and Krajbich [30].

[39] Loewenstein et al. [32], Konovalov and Krajbich [30].

[40] Kahneman [27].

[41] Weber and Johnson [63].

Brain imaging shows that in practically all decision-making tasks two regions of the brain are active—the ventromedial prefrontal cortex (vmPFC) and the striatum (encompassing basal ganglia), or at least some region of the striatum (e.g. the nucleus accumbens).[42] These regions seem to be involved in learning, storing, retrieving and updating values. Two things remain unclear: first, what exact computations take place in these regions; second, what the process of translating values into preferences and judgment is. What is clearer, is that those two regions are likely where the neural instantiation of utility takes place, with the posterior cingulate cortex most probably responsible for the numerical evaluation of magnitudes.[43] Higher probability of choosing a given option is associated both with activity in those regions and more intensive attention (gaze) directed at this option. This is the neurological explanation of the mere exposure effect. It seems that the neural computations in the brain occur mostly on differences and not on absolute values which is the neuroscientific explanation for the relative reference points that Prospect Theory identified.[44]

Neuroeconomics also studies how information is transmitted in the human brain. This takes place through the action potential (impulse or "spike") of the neurons. There is, however, a physical limit on this action potential—it is clearly a non-negative number but there is also an upper bound of roughly 100–200 spikes per second. As information is encoded in the rate of action potentials, the upper bound sets a limit of the range of values that can be usefully encoded. Subtle differences between e.g. probabilities will take a large amount of time to be represented and transmitted. To avoid this, the brain leverages adaptive normalization, making similar options look identical. In effect this leads to inability to perceive small differences between probabilities of the same order. Additionally, the inclusion of irrelevant alternatives decreases the differences between actual choices, thus leading to discrepancies and inconsistencies.[45] The former effect explains the decision-maker's indifference between close probabilities. The latter explains the breaches of the independence axiom of classical von Neumann-Morgenstern utility theory.

This result is one to give pause. It seems that the very structure of human's brain precludes completely rational decision-making, and thus it is by default that one needs to take recourse to decision aids for complex and high-stakes situations. Apart from the structural constraints of the brain, neuroeconomics has found that the mere volume of gray matter may have important repercussions for choice under risk and uncertainty. For example, a strong correlation has been found between the gray matter volume in the posterior parietal cortex (a brain region connected to risk preferences) and the level of risk tolerance found in individuals.[46] The argument here is that an increase in gray matter volume increases the computational capacity of the brain, leading individuals to estimate probabilities and outcomes more precisely,

---

[42] See Konovalov and Krajbich [30] and the references contained therein.

[43] Clithero and Rangel [15] and Kanayet et al. [29].

[44] Konovalov and Krajbich [30].

[45] Chau et al. [14].

[46] Guilaie-Dotan et al. [24].

thus loosening their risk constraints that in turn leads to a higher risk tolerance.[47] Again, the neuroanatomy of the brain puts physiological constraints on risk preferences and choices under uncertainty. Additionally, levels of certain hormones such as testosterone have been found to also be highly correlated with risk preferences, thus providing an even lower-level biochemical explanation of risk tolerance.[48] Such results give credence to using a highly personalized risk model when accounting for individual perceptions and decisions. The analyst should note, however, that this need not carry over to the aggregate level. It may very well be possible that a simpler model is able to capture the aggregated dynamics in a satisfactory manner.

Finally, an intriguing strand in neuroeconomic research has focused on predicting individual behavior under risk and identifying potential neurological drivers for group-level behavior. Herd behavior and bubble formation (consistent asset overvaluation) is a prime example that has puzzled risk analysts for decades. It sems that the neural activity of humans is strongly associated with market dynamics and irrational bubble formation. Using fMRI Smith et al.[49] showed that neural activity within the nucleus incumbens, NAcc (part of the oft-activated striatum) is in fact strongly correlated with overvaluation and risky behavior. In fact, NAcc activity may be used as a leading indicator to predict market crashes. A possible explanation may be that traders' activity is driven by their neural dynamics and some are able to intuitively expect crashes, thus changing their behavior which leads to a market correction. At any rate, this experiment clearly showed that investors who buy more aggressively due to NAcc activity tend to earn less. Higher earners, in contrast, experience early warnings and are able to reduce risk-taking behavior shortly before the market peak. They decrease their risk exposure and are thus able to realize better profits by buying low, selling high at the peak, and refraining from risk during the correction. The exact causal links behind those results are still subject to research but they clearly point at the importance of neural activity for determining risk exposure.

Using neuroeconomic approaches to model risk preference is a markedly different from leveraging standard approaches. Utility theory imposes a normative perspective on how people ought to make decisions, behavioral economics tests how people actually make decisions, and neuroeconomics studies the deep neurological low-level characteristics of the process. Unsatisfactory prediction by standard methods has sparked intense interest in understanding decisions by modeling them at the neural level, and there are some early successes.[50] Despite this, neuroeconomics has not yet produced a coherent consensus model for making choices under risk and uncertainty that can rival existing traditional and behavioral utility formulations. Thus, the overwhelming majority of researchers and analysts still use classical utility and Prospect Theory models taking into account both their potential limitations and the neuroanatomical explanations of some of their properties.

---

47 Woodford [64].

48 Apicella et al. [4].

49 Smith et al. [51].

50 Mengov [36].

6.5 Mediating Online Environment

Putting the individual into an online environment does little to ameliorate their biases and improve risk perceptions and probability evaluations towards some optimal standard. If anything, irrational behavior is still observed, and it has significant economic implications.[51] Decision-makers seem to conceptualize their online behavior using a separate mental account and are thus more prone to excessive, impulsive or bundled consumption. The richness of data on the internet gives rises to potential information processing problems such as information overload. In view of agents' finite cognitive resources this engenders the need for simple heuristics that in turn eat away at the optimality of choices made. Finally, offline phenomena such as herd behavior are greatly amplified on the internet, particularly in the context of online social media platforms. In short, the promised land of information and connectivity that the internet is sometimes touted to be, turns out to hold many traps for rationality.

The first set of moderating effects on individuals contains the different facets and characteristics of the online environment that includes web site or application quality, interface, satisfaction, and overall user experience.[52] Those need to be designed in such a way as to induce trust in the decision-maker with research showing that higher levels of trust are strongly associated with undertaking more risky behavior online.[53] In the context of internet activities, risky behavior is usually defined as being exposed to security or privacy risks, and it remains uncertain whether individuals have a well-calibrated estimate of the potential risks they tend to expose themselves to. Early inconsistencies in individual estimates show that the digital environment is particularly prone to inducing biases in probability estimates and thus—risk perception.[54] What is more, risks that have a physical analogue tend to be much better understood in an online environment while those that do not are more vaguely conceived.

The network nature of the digital environment also underlines a structure of influence, dependent on network positioning that can have significant effects on individuals, their perceptions, and resulting choices.[55] In this respect, well-positioned and active users tend to have an outsized effect on shaping the perception of members of their network. Paradoxically, user credibility or performance on relevant tasks may have no effect on their levels of influence.[56] The overwhelming focus on the quantity of content online may well come at the cost of evaluating its quality. Thus, an active network user with high centrality could significantly skew the perceptions and change the behavior of other users, even though the objective basis of this (e.g. expertise, credibility, results) is very weak. This phenomenon is amplified by another distinct feature of online communities—herding (herd behavior) whereby many individual take cue for their actions and perceptions from the overall group disposition. In an era

[51] Xie [66].

[52] Darley et al. [17].

[53] Chang and Fang [13], Wang et al. [62].

[54] Garg and Camp [21].

[55] Papagelis et al. [44].

[56] Ibid.

of increased connectivity and ever more customized connection recommendations, herding can be a substantial and powerful phenomenon, driven by two main factors: imitating others, and discounting own information.[57] A major insight is that herding may occur not only based on strong links (well-known individuals) but also on weak links which significantly magnifies its effect and increases the risks stemming from it.

The digital age has brought about information-rich online environments. The human brain is, however, poorly equipped to process all this data in a meaningful way and is thus susceptible to potentially deleterious information overload. This is hardly a new insight and it those concerns date back to at least Herbert Simon's ideas of bounded rationality. These state that the individual, being unable to process all the relevant information, expends effort to collect and leverage merely enough knowledge for a satisficing (good enough) decision. The exponential explosion of data in the internet has made it even more difficult to filter and evaluate, thus forcing individuals to find a number of coping strategies. A firmly established insight is that the phenomenon of information overload is tightly connected to psychological ill-being (including depressive and anger symptoms) and may lead individuals to scale down or discontinue search, retrieval and processing of potentially relevant insight.[58] What is more, the amount of information overload and information disorganization leads to a higher level of perceived risk by decision-makers.[59]

Human decision-makers tend to particularly avoid uncertainty and extensive amounts of data, or even excessive social interactions (i.e. social overload). This avoidance can be manifested in concrete action as when individuals are faced with information and social overload, they tend to discontinue the use of given online environments and services.[60] Such tendencies may lead to contradictory and surprising behavior. Information overload leads the users of knowledge repositories to stop using the resources therein, even though those resources are specifically curated to provide those same users with desired and relevant information.[61] Instead of relying on rational evaluation of risk and return when making decisions, individuals tend to get lost in the ocean of information online and thus fall back on a number of heuristics and biases that enable them to navigate (imperfectly) the online landscape.

There are some basic heuristics that human subject fall prey to when evaluating online information. Some of the more salient among them are as follows[62]:

- *Reputation heuristic*—it refers to the tendency for people to be more trustworthy and have higher evaluation of alternatives with established brands, names, or reputations. It essentially transfers the credibility of the brand upon individual pieces of information or recommendation, thus freeing the decision-maker from the need

57 Mattke et al. [33].

58 Swar et al. [54].

59 Soto-Acosta et al. [52].

60 Zhang et al. [67].

61 Bock et al. [9].

62 Metzger et al. [37].

to individually process and evaluate every piece of content. It is likely rooted in two basic features of human behavior: first, the tendency to favor recognized over unknown alternatives; second, the operation of the authority heuristic. Either way, this presents individuals with the opportunity to make decisions easier but also may engender suboptimal ones.

- *Endorsement heuristic*—individuals tend to be more trustful and less risk averse for alternatives that have been endorsed by somebody else, thus conferring the individual authority of the endorser on the option at hand. It is argued that it is group endorsement that has more powerful effects than the individual one. This is the well-known herding behavior, which is rooted in the manifestation of the consensus (or bandwagon) heuristic.
- *Consistency heuristic*—a common way to perform a quick and efficient information and choice validation is for individuals to ensure that there is consistency about this issue across a wider variety of sources. This is usually done on a relatively small set of alternative outlets and done in a quick and perfunctory way. Encouragingly, decision-makers tend to perform more thorough cross-validation when decisions involve high-stakes outcomes (e.g. financial transactions) or risky personal decisions (e.g. health situations). More worryingly, individuals find that online information is equally credible, if not more credible, than offline information.
- *Expectancy violation heuristic*—this heuristic operates in a more insidious ways and has the potential to bias and skew decisions towards irrationality. Individuals who employ it check whether the appearance (layout, features, functionality, etc.) or the content of a given sources corresponds to their expectations of what it *should* be. If this is not the case—the information or source is discarded as not credible. When the information is judged upon superficial characteristics such as presentation this may be the application of the liking heuristic that is applied not to a person but to an object. When the information is discarded on emotional grounds it is probably a manifestation of confirmation bias.
- *Persuasive intent heuristic*—this heuristic is often associated with the presence of commercial information whereby the source may be willing to persuade individuals to make a purchase or undertake a certain course of action. Decision-makers are irritated by this and tend to discard the information presented as not credible or not useful without the additional effort of inspecting it. Knowing this to be the case, persuasive intent sources tend to obfuscate this in order to better achieve their goals. It should be noted that the information presented need to have persuasive intent but only has to be perceived as such to invoke the heuristic.

While making choices in a complex online environment has numerous layers of complexity, it seems that individuals are ill-suited to process all the incoming information and thus form an unbiased probability judgment. If anything, exposure to internet content may further skew and radicalize risk perceptions, thus removing the decision-maker even further from optimality.[63] The risk analyst is thus called

[63] Anderson et al. [3].

upon to make proper corrections to the judgments elicited, or to change their context in order to ensure greater accuracy.

6.6 Information and Influence

As we have seen, the original sin of rational choice decision theory is overreliance on plausible assumptions rather than rigorous derivation of models from empirical data. While decisions in the lab are arguably different from real life situations, their analysis may still be able to bridge the gap between theory and data. This section presents insights from an economic experiment with 258 participants that are asked to make choices in situations of radical uncertainty. The experiment explores what factors influence their deliberations, what performance is achieved under varying amounts of information and uncertainty, and, crucially, what is the mediating effect of being exposed to social feedback through a simple online social network. The setup is simple—the experiment is framed as an economic game whereby participants are exposed to four suppliers that promise the delivery of a certain amount of a generic good (omnium bonum[64]). Subjects then select their preferred supplier and receive information of the actual amount of good delivered, which is often different from the promise.[65] Then participants evaluate their satisfaction with the selected supplier and are asked to do this again over a total of twenty rounds of the game. This yields a total of 5,160 decisions under risk that can be further analyzed.

Some of the subjects operate in an economy that only grows, meaning that the total amount of the good that could be delivered (a synthetic measure for GDP) is increasing throughout the rounds, while others are faced with a cyclic economy of growth followed by a decline. Half of the participants were exposed to information about the GDP growth, while the other half was not. Finally, half of the subjects could see what evaluations others gave of the supplier, thus mimicking a simple online social network; the other half were on their own. In short, the experiment follows a classical randomized $2 \times 2 \times 2$ design. Participants had no idea of the underlying dynamics of the game putting them in a situation of radical uncertainty. However, the experimental design was carefully crafted so that the volatility of delivery was directly proportional to the delivered amount, which reflects the well-known connection between risk and expected return (see Fig. 6.5).

6.6.1 *Humans as Intuitive Statisticians*

The view about human decision-makers is vastly different in economics as compared to psychology. In the former, much emphasis is put on systematic deviations from

[64] From Latin—"a good for everyone".

[65] Results from the first smaller-scale experiment with this setup are reported in [34].

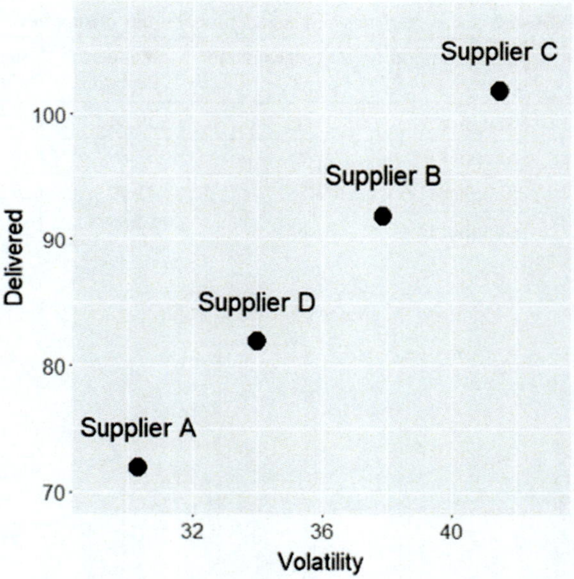

Fig. 6.5 Mean and standard deviation of actual [delivery] of omnium bonum in the economic experiment

rationality, whereas in the latter the focus is on the fact that individuals are still able to make imperfect decisions that are mostly "good enough". This gave rise to the view of humans as intuitive Bayesian statisticians. Essentially, this idea boils down to the fact that decision-makers in situations of risk and uncertainty are able to form approximate estimations of relevant probabilities and learn as feedback is given to them. In the experimental game, people are put in a situation of radical uncertainty—they have no idea of the underlying economic dynamics, the reliability of suppliers, and the generating process behind actual deliveries. Yet, their choices reflect an amazing intuitive understanding of the risky situation.

The design behind the game is such that the greatest expected delivery is provided by Supplier C, followed by Supplier B, and then by Supplier D and Supplier A, respectively (see Fig. 6.5). This ordering is exactly replicated in the average choice probabilities. In short, individuals are most likely to choose the highest delivery supplier and least likely to choose the lowest delivery supplier even though they have no idea that this is the case. It seems that participants learn to evaluate the expected risks and outcomes over the game through its feedback mechanism and end up with rather precise estimates of what to expect: the average choice probability of each supplier is almost identical to the share of total output of this supplier (Table 6.2). It seems that there is also an effect of risk aversion—the supplier with the lowest standard deviation (level of risk), Supplier A, is selected much more often that its expected payoff would imply (Fig. 6.6).

The remarkable ability to form intuitive judgment in the experiment also carries over in experimental conditions that have different underlying dynamics. Participants are able to identify and select highest performing suppliers in both situations with constant output growth, and in situations that have initial growth, followed by a

Table 6.2 Characteristics and choice probabilities of suppliers

#	Choice option	Average amount	Standard deviation	Share of total amount (%)	Average choice probability (%)
1	Supplier A	71.88	30.34	20.7	17.1
2	Supplier B	91.88	37.89	26.4	28.3
3	Supplier C	101.88	41.49	29.3	33.4
4	Supplier D	81.88	33.99	23.6	21.2

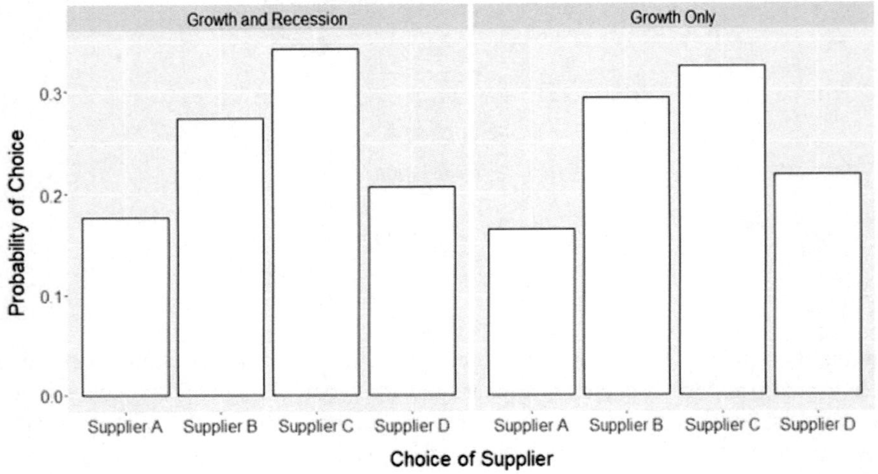

Fig. 6.6 Probability of choosing a given supplier across two economic conditions

decline, thus mimicking the economic cycle. Under both conditions, the ordering of the choice probabilities remains the same and it almost perfectly reflect the expected delivery—a quantity that participants never observe. If anything, the cyclic conditions accentuates choice with a slight spillover from Supplier B to Supplier C, thus showing that more participants correctly identify Supplier B as a somewhat dominated alternative in term of both risk and reward.

6.6.2 Performance Under Uncertainty

One of the key questions when studying risk preferences is the eventual performance of individuals as they try to navigate an uncertain world (or experimental problem). Armed with an intuitive ability to relate risks and outcomes, participants are able to better maximize their payoff. They are also incentivized to do, as at the end of the experiment the amount of fictional good is translated into actual money they go home with. The challenge here is that the amount that is going to be delivered by any supplier is unknown to the participants and thus they must optimize upon a reconstructed estimate (or guess) of expectation and risk. Despite the very limited available information, individuals are still able, on average, to pick suppliers with higher delivered amounts (Fig. 6.7). The probability of choosing a given supplier in a given round is directly proportional to the amount to be delivered in this round. Since this future amount is a reflection of the overall expected delivery, it seems that players form a relatively good estimate of this quantity. This result holds across both modes of operation of the experimental economy—growth only; and growth and recession.

While this holds as a general trend, there are also many deviations that show that while decisions are good enough, they are far from perfect and sometimes large discrepancies and gaps between optimality and reality can be observed. This can be further investigated in a linear regression framework, regressing choice probability on delivery. The coefficient on delivery is highly statistically significant ($p < 0.005$) but the total explained variance is relatively small with adjusted $R^2 = 0.1$.

This demonstrates that while intuition and learning are important drivers of performance in uncertain situations, they still remain an imperfect substitute for more sophisticated methods and improved information. The particularities of human perception can be further investigated by reviewing the link between performance and choice probabilities at the level of the individual supplier (Fig. 6.8). There is a clear link between amount to be delivered and the probability of choosing the supplier for Supplier A and Supplier C—it is visually detectable, and it reaches statistical significance at levels below 1%. On the other hand, the connected between amount delivered and probability of choosing the supplier all but breaks down for Supplier B and D.

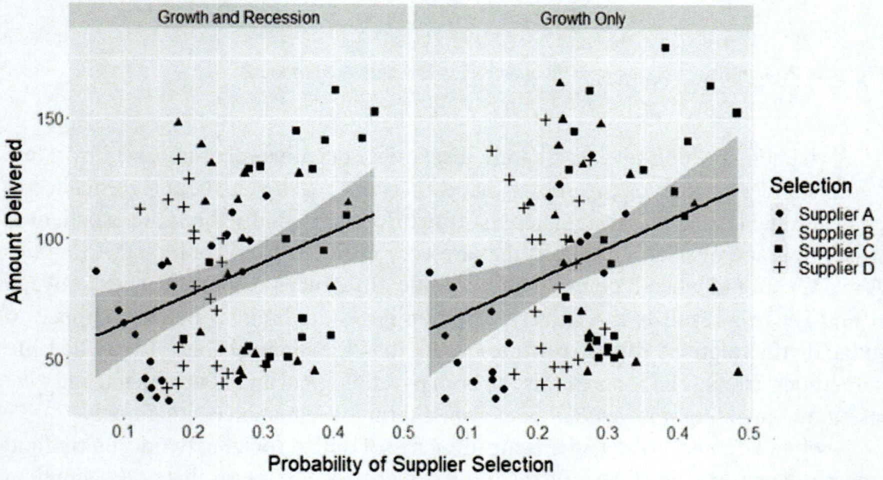

Fig. 6.7 Probability of choosing a given supplier and delivered amount across conditions

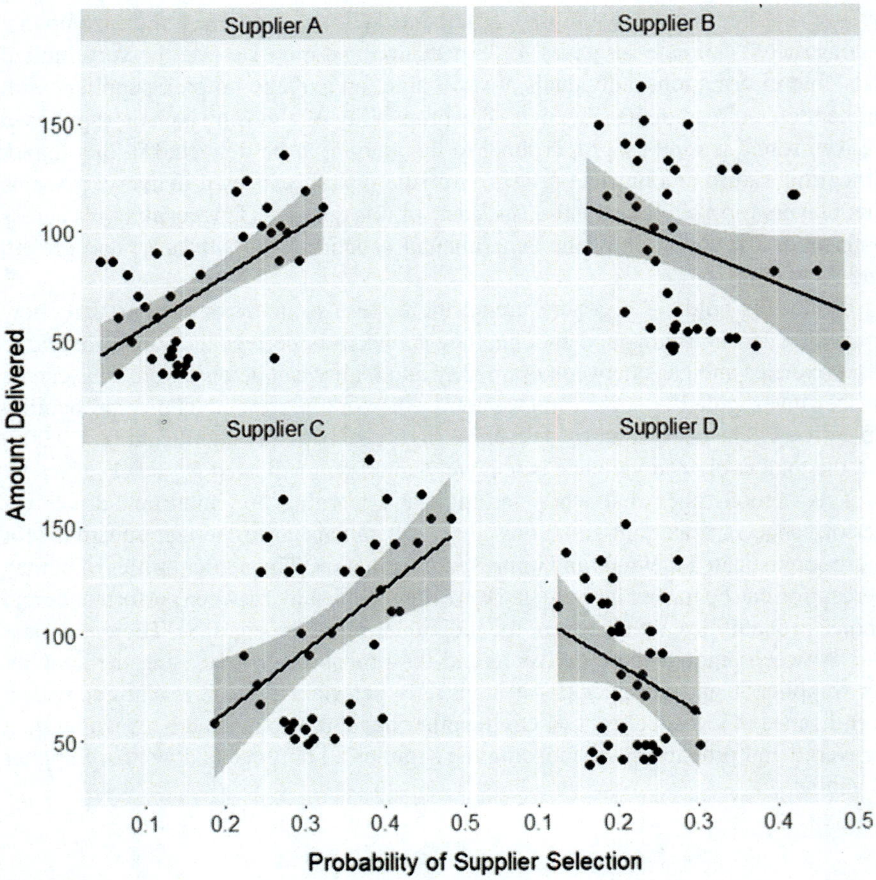

Fig. 6.8 Probability of choosing a given supplier and delivered amount

A visual inspection reveals a barely discernible negative relation driven by a few outliers and a formal testing shows that it does not reach statistical significance at conventional levels. The insight here is straightforward—individuals are excellent at discerning and evaluating options that are very different from each other (Supplier A and C) but fail when it comes to more subtle differences such as the ones between B and D. This result is a relatively common one—the brain is better equipped to judge distinct options (black or white) rather than the many shades of gray that life sometimes offers. The roots of this are both psychological and neurological and give rise to the puzzling relevance of irrelevant alternatives in choice situations.

Another key issue to consider is the influence of online social networks on risk and return perceptions, and thus—on resulting performance. It seems that under identical conditions, humans perform better when they have access to a forecast about future output, which proxies relevant economic aggregates in real-life decision situations (Fig. 6.9). In fact, participant could not possibly know that this is the perfect forecast

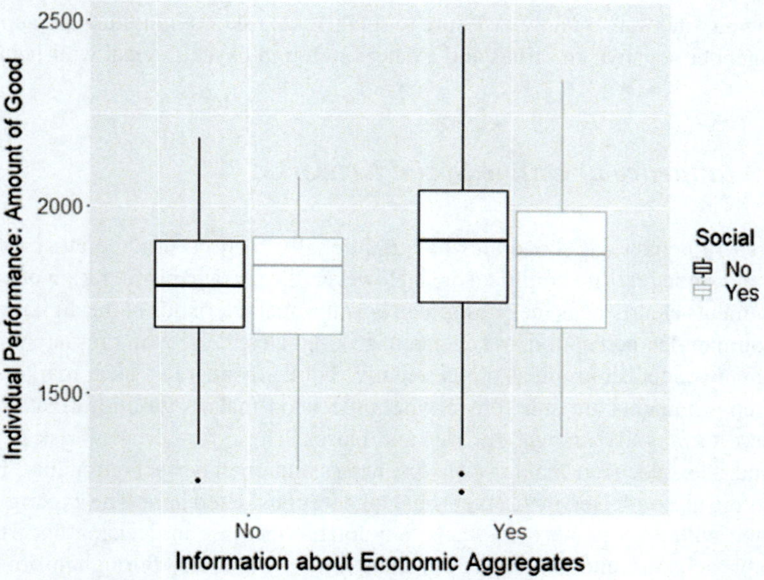

Fig. 6.9 Probability of choosing a given supplier in conditions with or without a social network

for the next round, as it merely displays the sum of total deliveries for the next round according to the experimental plan. Yet, the very presence of informational stimuli makes decision-makers more rational and strategic, prompting the decision maker to choose different suppliers ($\chi^2 = 25.59, p < 0.005$). This in turn leads to an increase in cumulative results. This difference is practically relevant with an average increase of 80 units and reaches statistical significance at levels much below 1% (Table 6.3).

On the other hand, participation in a social network during the game has much more mixed results. Where no objective economic information is given, i.e. the conditions without aggregates, exposure to real time online social cues leads to a slightly improved performance. In the presence of the output forecast, the social network leads to a deterioration in performance. Overall, the effects on resulting performance are rather small in magnitude and fail to reach significance. In contrast, participant satisfaction is lower in the social network conditions, with this effect being highly significant. In short, while people do not do worse in absolute terms, the constant social comparisons diminish their utility. This result echoes the insights from

Table 6.3 Difference in outcomes depending on experimental condition

	Experimental conditions	Performance			Satisfaction		
		F-stat	Sig	Delta	F-stat	Sig	Delta
1	Growth or cycle	31.77	<0.005	−138.2	0.218	0.641	−0.03
2	Economic information or no data	9.896	0.0019	80.3	0.13	0.718	0.03
3	Social network or individual	0.31	0.578	−14.46	6.946	0.008	−0.197

the growing literature on how exposure to social media skews individual perceptions and engender negative emotions and a deterioration in psychological well-being.[66]

6.6.3 *Influence of Online Social Networks*

Research in behavioral economics has traditionally focused on laboratory experiments and some real-life empirical data. However, the rise and proliferation of online environments clearly begs the question on how the characteristics of the digital world affect human risk perception and decision-making. Despite growing research on the topic, many conclusions are as of yet tentative but there are a few clear markers that online environments are able to overwhelm decision-makers with information and can serve to magnify known attitudes and biases. These further skew risk perceptions and steers decision-making away from the standard of perfect rationality. Those insights are also reflected in the experimental data considered here. The experimental condition with a social network leads to a notable increase in decision uncertainty for the participants, and if needed this can be formally modeled through information theory concepts such as entropy.[67]

Some of the participants in this economic game were exposed to information about other people's satisfaction with providers, thus simulating a simple version of an online social network. Such approach mimics the dynamics of opinion sharing and engenders phenomena such as supplier reputation (or average user satisfaction) as seen by a target group (i.e., a market segment). To better understand the underlying dynamics, resulting choices are modeled via a neural network.[68] The relative importance of decision factors can be seen in Fig. 6.10, and the results are striking.

First, reputational effects turn out to be among the most important decision-predictors in conditions with a social network. This shows that group dynamics and herd behavior have important repercussions on economic decisions. The prominence of the reputation of Suppliers B and D is of particular note here—both are somewhat dominated choices that offer neither the highest return, nor the lowest risk. Participants are able to form intuitive evaluations of this and communicate it through to others in their social networks via means of a very sparse signal—a sequence of satisfaction scores. Such results give credence to real-life rating systems whereby users evaluate a supplier, good, or experience through a simple scale (e.g. one to five stars) and show that such mechanisms can be useful and definitely have a powerful effect on perception and choice.

Second, the amount of effort put into the decision seems to be higher when participants are exposed to the social network. While the time expended on deliberation is relatively less important under the social condition as there are other factors that

---

66 Boroon et al. [10], Villalonga-Olives and Kawachi [59].

67 Atanasov [6].

68 Iterative modeling showed that a neural network with three hidden layers exhibits both good fit and adequate parsimony.

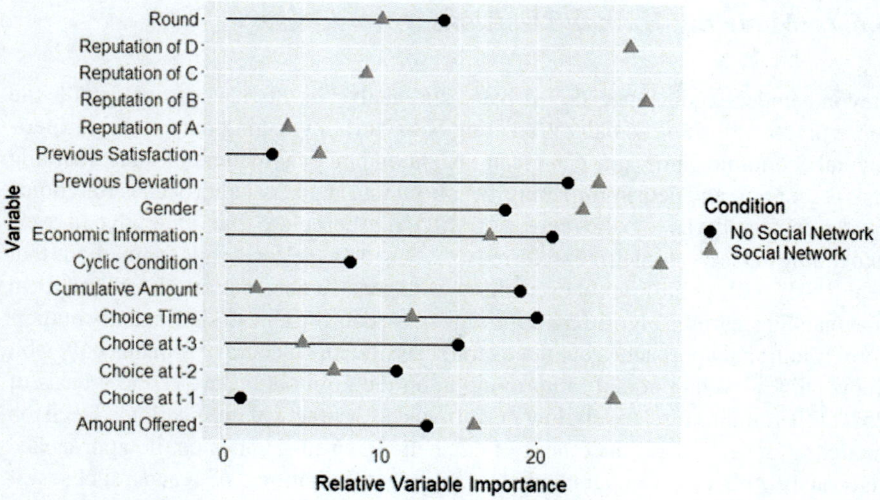

Fig. 6.10 Relative predictor importance across two economic conditions

weight in, participants spend much more time on determining their choice and the difference is highly statistically significant ($t = -8.21, p < 0.005$). It seems that being exposed to more information leads to longer decision times but there is practically no improvement in the quality of the decision—there is no difference in final outcomes between social and individual conditions (see Table 6.3) even though people are less satisfied with their decision. In short, the social network makes a decision more difficult to make by providing useless information and participants are not happier with it.

Third, choices become more myopic, focusing on immediately preceding action or emotion, thus diminishing the importance of longer time series. The previous deviation between offer and delivery, the directly preceding choice, and the last felt satisfaction increase in importance. At the same time, choices in preceding rounds ($t - 2$ and $t - 3$) markedly drop in importance. The social network amplifies the importance of proximate experience, while simultaneously amplifying the sensitivity to change. The latter is visible in the increased prominence of the cyclic condition.

Fourth, it seems that participants move the burden of choice from their own deliberation to the wisdom of the network. Under the social condition, typically important individual-level factors such as total personal outcomes (proxied by cumulative result), choice consistency (proxied by lagged choices), gained experience (proxied by round number) are not as important predictors. Against this backdrop, all network-related information is gaining prominence—from the average reputation (pictured in Fig. 6.10), through current round real-time evaluations, through diverse combinations of both of those. Essentially, when put in a social network, decision-makers count less on themselves, and more on the network. Such behavior clearly gives rise to herding and other familiar group dynamics in the online environment.

6.6.4 Modeling Human Decisions

Beyond understanding decision drivers, the analyst is often confronted with the more practical task of actually predicting behavior. Leveraging data on this experimental economic game, one can glean insight on practical matters of understanding risk perception and decision-making. While classic statistical approaches for choice modeling such as the logistic regression still have their place, the imperative of ever-expanding datasets (sometimes referred to as "big data") clearly calls for methods that seamlessly scale to millions or even billions of agents or observations. It can be shown that machine learning algorithms applied to these data are able to significantly outperform traditional approaches but the overall classification accuracy remains only a bit above 40%.[69] While notably improving upon the no-model (pure chance) basis of 25%, this remains a relatively low performance for practical applications. Decision modeling under radical uncertainty is a challenging task and must be approached rigorously with a clear understanding of pertinent limitations. As a general observation, novel machine learning methods as well as neural modeling approaches tend to outperform simple statistical algorithms.[70]

To further explore this insight, a set of fifty-five of the most popular advanced machine learning (ML) algorithms are applied to the data at hand. The models are trained on 80% of the data (train set) and then applied to the other 20% (test set) to avoid issues of overfitting. Their classification accuracy is compared through the area under the Receiver Operating Characteristic (ROC) curve and presented graphically in Fig. 6.11. The area under the ROC curve (AUROC metric) presents a balanced view of classification accuracy, taking into account both the true and the false positive rates. Realistic AUROC ranges are from 0.5 to 1, whereby 0.5 is equal to pure chance random classification, while 1 corresponds to perfect classification. AUROC values that fall below 0.5 indicate a model that performs worse than random selection.

The algorithm review in Fig. 6.11 shows instructive results with alternative algorithms having very different performance metrics. A lot of algorithms in both the no network and the social network conditions fail to meaningfully distinguish between choices—the peak in the AUROC distribution under both conditions is at 0.5. The are some methods with mediocre performance, and a few that show excellent performance. The analyst must therefore make a judicious choice when modeling decisions under risk and select an appropriate high-performance tool. This will likely vary across different situations in reflection of the No Free Lunch Theorem[71] and thus a range of alternatives needs to be tested before settling on the final one.

The best performing algorithms in this particular problem seem to be versions of linear models, modified logistic regressions, and partial least squares approaches (see Table 6.4). The family of Classification and Regression Trees (CART) consistently

---

[69] See Gerunov [22].

[70] Gerunov [22], Mengov [35].

[71] The basic idea is that algorithms that have higher performance on a specific task tend to underperform on other tasks, thus there is no single best algorithm or solution. A recent review is found in [1].

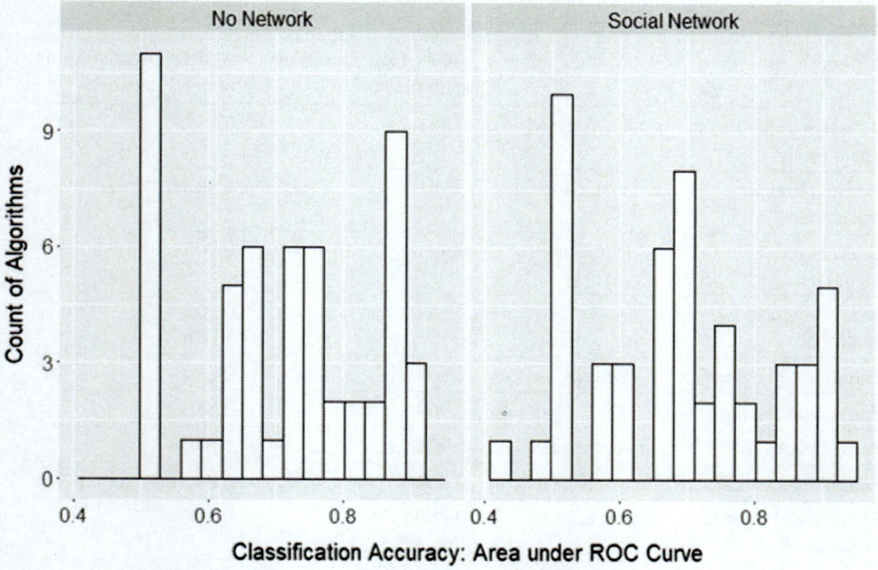

Fig. 6.11 Performance of machine learning algorithms for choice modeling

showed overfitting in this problem with AUROC = 1, even in the test set but this is a problem that can be avoided by increasing the sample size or changing the data processing approach. The overall conclusion from this study and other extant research[72] is that specific algorithmic families tend to perform better, on average, in different classification problems:

- **Classical statistics** (e.g. logistic regression) tends to perform well in small sample sizes (hundreds or thousands of observations) with a relatively limited number of predictors. In this case the levels of statistical significance remain relatively reliable and the coefficient sizes have a meaningful interpretation.
- **Random Forest models** (ensembles of CART) performs excellently when the sample sizes are large (between thousands and hundreds of thousands observations) and the number of predictors is relatively limited (tens to hundreds). While the individual tree is extremely prone to overfitting, the ensembling of a sufficiently large number of trees does much to improve upon this. At this point, levels of statistical significance and coefficient sizes are less meaningful and the analyst may instead want to focus on improvement of classification accuracy and relative variable importance measures (such as reduction in the Gini impurity coefficient).
- **Neural Networks** —they seem to work best as the data size and the number of predictors explodes from tens of millions to billions of observations with thousands or more explanatory variables. Improvements upon plain vanilla neural networks such as the introduction of reinforcement learning, or the

[72] Fernandez-Delgado et al. [20].

Table 6.4 Performance of best fitting modeling algorithms

Algorithm family	Method name	AUROC in conditions with network	AUROC in conditions with no network
Bayesian generalized linear model	bayesglm	0.853	0.835
Boosted logistic regression	LogitBoost	0.951	0.882
Generalized additive model using LOESS	gamLoess	0.904	0.906
Nearest shrunken centroids	pam	0.887	0.908
Partial least squares	simpls	0.904	0.880
Partial least squares	pls	0.904	0.880
Partial least squares	kernelpls	0.904	0.880
Partial least squares	widekernelpls	0.904	0.880
Regularized logistic regression	regLogistic	0.880	0.884
Single rule classification	OneR	0.721	0.908
Support vector machines with polynomial kernel	svmPoly	0.863	0.885

generative adversarial networks improve their performance further. Generally, neural networks are the underlying technology for most applied domain-specific Artificial Intelligence and are widely used in more complex settings.

While those insights present a very crude rules of thumb, they may still be useful as a rough starting point for model exploration.

The last remaining question is how the digital environment affects people's perceptions and decisions, and whether this has any impact on the analyst's ability to formally model those. Figure 6.12 presents the smoothed algorithms distribution given their classification accuracy across the two conditions. The difference is obvious—overall the condition without social network provides an environment in which decisions are easier to forecast. Both the average AUROC, and the AUROC of the best performers is higher as compared to the condition with social network. It seems that the introduction of social network unlocks a number of decision drivers such as social comparison, herding, and information overload that make behavior less predictable. It does not change the type of best performing algorithms (i.e. they are almost the same, see Table 6.4) but decreases their predictive ability.

In short, the more data becomes available to the decision-maker, and the more exposure to online social interaction occurs, the harder it is to precisely model risk perceptions and judgment. The digital environment is a ball game of its own and it significantly affects individual thinking and behavior.

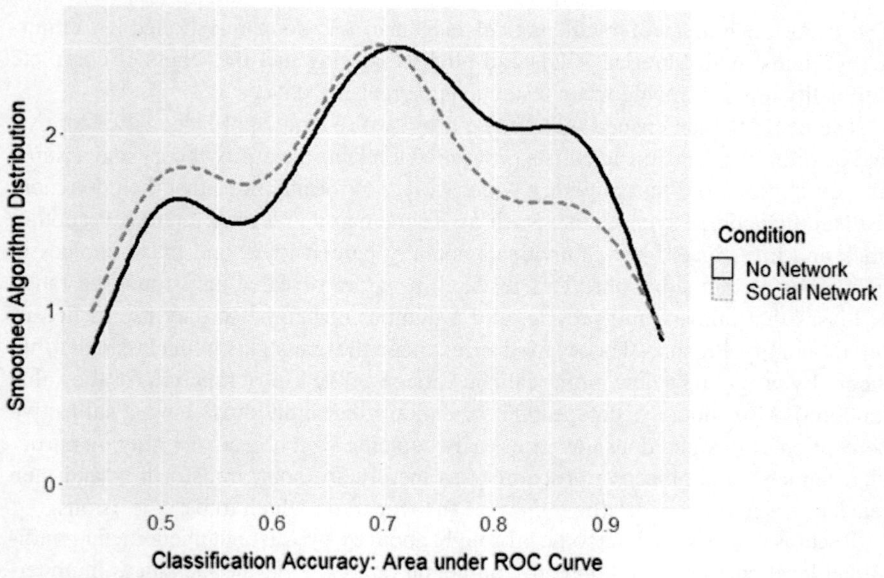

Fig. 6.12 Aggregate performance in conditions with social network (solid line) and without social network (dashed line)

6.7 Conclusion

Understanding risk is a daunting endeavor. The human brain is evolutionary hard-wired to process and make judgments in situations of uncertainty. The major challenge is that the environment of humans has changed much more rapidly than our brains, thus putting modern decision-makers in some situations of risk, for which they are poorly prepared. Suboptimal choices may follow as a result. The risk analyst then has a two-pronged task: first, to understand how people and organizations view and conceptualize risk in order to support a risk-adjusted decision; and second, to model and possibly predict choice in situations of uncertainty. While the arsenal of tools has rapidly expanded to take advantage of exponential data growth and advanced machine learning algorithms, the increase in the complexity of modeling has more than kept apace.

Early decision-theorists approached the challenge of modeling risk by relying on simple parsimonious models of human choice as encoded in tractable and well-behaved mathematical formulations of preferences—the so-called "utility functions". While a useful first pass at modeling, those utility functions imposed a number of wildly unrealistic assumptions such as stationarity of preference over time, clarity and transitivity of choices, and completely rational assessment of alternatives. The high point in these ideas was reached as von Neuman and Morgenstern completed their 1954 magnum opus *Theory of Economic Games and Economic Behavior.*

The beautiful edifice of mathematical elegance was soon questioned on empirical grounds, with Maurice Allais and others exposing that the tenets of complete rationality are not tenable when tested against real-life data.

The decades than ensued were marked by two distinct approaches. Economists and quantitative decision scientists proceeded to elaborate utility theory and assume new utility functions that provide a wider variety of formulations that included more realistic behavior. Psychologists such as Kahneman and Tversky pursued another track and investigated how individuals actually perceive risk and make choices in risky or uncertain situations. This research program revealed an astounding range of biases that humans fall prey to, and a number of heuristics they use in lieu of full rationality. Prospect Theory tried to reconcile those insights with classical utility theory by providing a new more realistic version of the utility function (or the value function). Most notably, Prospect Theory shows that individuals have a subjective perception of risk that does not necessarily coincide with objective reality. In particular, humans take objective probability as merely an input, transform it, and then reach a subjective number (how *they* see the world) on which to base decisions.

It seems that this final outcome is brought about by two distinct phenomena—individual level characteristics and environmental factors.[73] Further advances in understanding the neuroanatomy of the brain aim to address the first set of factors on a much deeper level. While the field of neuroeconomics is relatively new, results indicate that both physical characteristics (amount of gray matter, spike capacity of neurons) as well as structure and connections impinge on human decisions. More particularly, it is confirmed that uncertainty is assessed as specific parts of the brain (e.g. the striatum) get activated and a larger amount of gray matters corresponds to higher risk tolerance due to increased computational capacity. Physical limits on neural transmission, on the other hand, hinder the communication of very nuanced probability estimates, thus blurring risk evaluation. Hormonal levels also play a role in risk attitude, with testosterone being correlated with risk-taking. The nature of the individual risk preference can then be analyzed from the lowest level—brain activity, through abstractions such as the System I and II all the way into resulting cognition, which is traditionally the focus of behavioral economists and decision scientists.

The environment around the individual also has important implications for the risk attitudes and resulting judgment. There is an increasing realization that an appropriate decision architecture will make some risks more or less salient and thus influence (or nudge) choice. At the level of environmental influence a number of emergent factors surface. Most notably, choice presentation, information overload and social influence have important repercussions on how people make decisions. The proliferation of online environments such as online social networks puts humans in yet another unfamiliar environment where they need to evaluate risk and make a choice. Early insights show that individuals are particularly susceptible to information overload whereby they suspend their complex rational judgment in favor of simple heuristics. Additionally, herd behavior and group thinking are salient drivers of individual decisions. In the simple economic game that was analyzed in this chapter, one can clearly

[73] Those are known in popular discourse as "nature" and "nurture" factors, respectively.

observe that in a rudimentary online social network participants tend to undervalue their own experiences in favor of consensus network opinion.

The digital era has brought not only better tools to understand human risk perceptions and judgment but also exposed individuals to novel situations in which they may struggle to cope. The research challenge here is to better understand those effects in order to support optimal individual and organizational decisions. The practical challenge is to model risks in such a way that can bring about a meaningful and tangible improvement in individual satisfaction and social welfare.

References

1. Adam, S.P., Alexandropoulos, S.A.N., Pardalos, P.M., Vrahatis, M.N.: No Free Lunch Theorem: A Review. Approximation and Optimization, pp. 57–82 (2019)
2. Allais, M.: Le comportement de l'homme rationnel devant le risque: critique des postulats et axiomes de l'école Américaine. Econometrica **21**(4), 503–546 (1953)
3. Anderson, A.A., Brossard, D., Scheufele, D.A., Xenos, M.A., Ladwig, P.: The "nasty effect:" online incivility and risk perceptions of emerging technologies. J. Comput. Mediat. Commun. **19**(3), 373–387 (2014)
4. Apicella, C.L., Dreber, A., Campbell, B., Gray, P.B., Hoffman, M., Little, A.C.: Testosterone and financial risk preferences. Evol. Hum. Behav. **29**(6), 384–390 (2008)
5. Ariely, D., Jones, S.: Predictably Irrational. Harper Audio, New York, NY, (2008)
6. Atanasov, I.S.: Uncertainty in social networks: a matter of opinions. IFAC-PapersOnLine **52**(25), 494–497 (2019)
7. Babcock, B.A., Choi, E.K., Feinerman, E.: Risk and probability premiums for CARA utility functions. J. Agric. Resour. Econ. 17–24 (1993)
8. Barberis, N.C.: Thirty years of prospect theory in economics: a review and assessment. J. Econ. Perspect. **27**(1), 173–196 (2013)
9. Bock, G.W., Mahmood, M., Sharma, S., Kang, Y.J.: The impact of information overload and contribution overload on continued usage of electronic knowledge repositories. J. Organ. Comput. Electron. Commer. **20**(3), 257–278 (2010)
10. Boroon, L., Abedin, B., Erfani, S.: Exploring the Dark Side of Online Social Networks: A Taxonomy of Negative Effects on Users. Research-in-Progress Papers, 30. (2018)
11. Brocas, I., Carrillo, J.D., Giga, A., Zapatero, F.: Risk aversion in a dynamic asset allocation experiment. J. Financ. Quant. Anal. **54**(5), 2209–2232 (2019)
12. Camerer, C.F., Loewenstein, G., Prelec, D.: Neuroeconomics: why economics needs brains. Scand. J. Econ. **106**(3), 555–579 (2004)
13. Chang, Y.S., Fang, S.R.: Antecedents and distinctions between online trust and distrust: predicting high-and low-risk internet behaviors. J. Electron. Commer. Res. **14**(2), 149 (2013)
14. Chau, B.K.H., Kolling, N., Hunt, L.T., Walton, M.E., Rushworth, M.F.S.: A neural mechanism underlying failure of optimal choice with multiple alternatives. Nat. Neurosci. **17**(3), 463–470 (2014)
15. Clithero, J.A., Rangel, A.: Informatic parcellation of the network involved in the computation of subjective value. Soc. Cognit. Affect. Neurosci. **9**(9), 1289–1302 (2014)
16. Conniffe, D.: The flexible three parameter utility function. Ann. Econ. Financ. **8**(1) (2007)
17. Darley, W.K., Blankson, C., Luethge, D.J.: Toward an integrated framework for online consumer behavior and decision making process: a review. Psychol. Mark. **27**(2), 94–116 (2010)
18. Dowd, K., Cotter, J., Sorwar, G.: Spectral risk measures: properties and limitations. J. Financ. Serv. Res. **34**(1), 61–75 (2008)

19. Eisenhauer, J.G., Ventura, L.: Survey measures of risk aversion and prudence. Appl. Econ. **35**(13), 1477–1484 (2003)
20. Fernández-Delgado, M., Cernadas, E., Barro, S., Amorim, D.: Do we need hundreds of classifiers to solve real world classification problems? J. Mach. Learn. Res. **15**(1), 3133–3181 (2014)
21. Garg, V., Camp, J.: End user perception of online risk under uncertainty. In: 2012 45th Hawaii International Conference on System Sciences, pp. 3278–3287. IEEE (2012)
22. Gerunov, A.: Modelling economic choice under radical uncertainty: machine learning approaches. Int. J. Bus. Intell. Data Min. **14**(1–2), 238–253 (2019)
23. Gigerenzer, G.: The bias in behavioral economics. Rev. Behav. Econ. **5**(3–4), 303–336 (2018)
24. Gilaie-Dotan, S., Tymula, A., Cooper, N., Kable, J.W., Glimcher, P.W., Levy, I.: Neuroanatomy predicts individual risk attitudes. J. Neurosci. **34**(37), 12394–12401 (2014)
25. Guiso, L., Paiella, M.: Risk aversion, wealth, and background risk. J. Eur. Econ. Assoc. **6**(6), 1109–1150 (2008)
26. Harel, A., Francis, J.C., Harpaz, G.: Alternative utility functions: review, analysis and comparison. Rev. Quant. Financ. Acc. **51**(3), 785–811 (2018)
27. Kahneman, D.: Thinking, fast and slow. Macmillan (2011)
28. Kahneman, D., Tversky, A.: Prospect theory: an analysis of decision under risk. Econometrica **47**(2), 263–292 (1979)
29. Kanayet, F.J., Opfer, J.E., Cunningham, W.A.: The value of numbers in economic rewards. Psychol. Sci. **25**, 1534–1545 (2014)
30. Konovalov, A., Krajbich, I.: Over a decade of neuroeconomics: what have we learned? Organ. Res. Methods **22**(1), 148–173 (2019)
31. l'Haridon, O., Vieider, F.M.: All over the map: a worldwide comparison of risk preferences. Quant. Econ. **10**(1), 185–215 (2019)
32. Loewenstein, G., Rick, S., Cohen, J.D.: Neuroeconomics. Annu. Rev. Psychol. **59**, 647–672 (2008)
33. Mattke, J., Maier, C., Reis, L., Weitzel, T.: Herd behavior in social media: the role of Facebook likes, strength of ties, and expertise. Inf. Manag. **57**(8), 103370 (2020)
34. Mengov, G., Egbert, H., Pulov, S., Georgiev, K.: Emotional balances in experimental consumer choices. Neural Netw. **21**(9), 1213–1219 (2008)
35. Mengov, G.: Person-by-person prediction of intuitive economic choice. Neural Netw. **60**, 232–245 (2014)
36. Mengov, G.: Decision Science: A Human-Oriented Perspective. Springer, Berlin Heidelberg (2015)
37. Metzger, M.J., Flanagin, A.J., Medders, R.B.: Social and heuristic approaches to credibility evaluation online. J. Commun. **60**(3), 413–439 (2010)
38. Meyer, J.: Representing risk preferences in expected utility based decision models. Ann. Oper. Res. **176**(1), 179–190 (2010)
39. Mongin, P.: The Allais paradox: what it became, what it really was, what it now suggests to us. Econ. Philos. **35**, 423–459 (2019)
40. Montibeller, G., Von Winterfeldt, D.: Cognitive and motivational biases in decision and risk analysis. Risk Anal. **35**(7), 1230–1251 (2015)
41. Mousavi, S., Gigerenzer, G.: Risk, uncertainty, and heuristics. J. Bus. Res. **67**(8), 1671–1678 (2014)
42. Ñíguez, T.M., Paya, I., Peel, D., Perote, J.: On the stability of the constant relative risk aversion (CRRA) utility under high degrees of uncertainty. Econ. Lett. **115**(2), 244–248 (2012)
43. O'Donoghue, T., Somerville, J.: Modeling risk aversion in economics. J. Econ. Perspect. **32**(2), 91-114 (2018)
44. Papagelis, M., Murdock, V., van Zwol, R.: Individual behavior and social influence in online social systems. In: Proceedings of the 22nd ACM Conference on Hypertext and Hypermedia, pp. 241–250 (2011)
45. Perets, G.S., Yashiv, E.: The fundamental nature of HARA utility. CFM discussion paper series (CFM-DP2015-22). Centre For Macroeconomics, London, UK (2015)

46. Puaschunder, J.M.: Nudgitize Me! A behavioral finance approach to minimize losses and maximize profits from heuristics and biases. J. Org. Psychol. **18**(1) (2018)
47. Saha, A.: Flexible utility form. Am. J. Agric. Econ. **75**, 905–913 (1993)
48. Saposnik, G., Redelmeier, D., Ruff, C.C., Tobler, P.N.: Cognitive biases associated with medical decisions: a systematic review. BMC Med. Inform. Decis. Mak. **16**(1), 138 (2016)
49. Settlage, D.M., Preckel, P.V.: Robustness of non-parametric measurement of efficiency and risk aversion. In: 2002 Annual Meeting, July 28–31, Long Beach, CA (No. 19765). American Agricultural Economics Association (2002)
50. Simon, M., Houghton, S.M., Aquino, K.: Cognitive biases, risk perception, and venture formation: how individuals decide to start companies. J. Bus. Ventur. **15**(2), 113–134 (2000)
51. Smith, A., Lohrenz, T., King, J., Montague, P.R., Camerer, C.F.: Irrational exuberance and neural crash warning signals during endogenous experimental market bubbles. Proc. Natl. Acad. Sci. **111**(29), 10503–10508 (2014)
52. Soto-Acosta, P., Molina-Castillo, F.J., Lopez-Nicolas, C., Colomo-Palacios, R.: The effect of information overload and disorganisation on intention to purchase online. Online Inf. Rev. (2014)
53. Sunstein, C.R., Thaler, R.H.: Libertarian paternalism is not an oxymoron. Univ. Chicago Law Rev. 1159–1202 (2003)
54. Swar, B., Hameed, T., Reychav, I.: Information overload, psychological ill-being, and behavioral intention to continue online healthcare information search. Comput. Hum. Behav. **70**, 416–425 (2017)
55. Thaler, R.H., Sunstein, C.R.: Libertarian paternalism. Am. Econ. Rev. **93**(2), 175–179 (2003)
56. Thaler, R.H., Sunstein, C.R.: Nudge: Improving Decisions About Health, Wealth, and Happiness. Penguin (2009)
57. Toda, A.A.: Necessity of hyperbolic absolute risk aversion for the concavity of consumption functions (2020). arXiv preprint arXiv:2009.13564
58. Tversky, A., Kahneman, D.: Advances in prospect theory: cumulative representation of uncertainty. J. Risk Uncertain. **5**(4), 297–323 (1992)
59. Villalonga-Olives, E., Kawachi, I.: The dark side of social capital: a systematic review of the negative health effects of social capital. Soc. Sci. Med. **194**, 105–127 (2017)
60. von Neumann, J., Morgenstern, O.: Theory of Economic Games and Economic Behavior. Princeton University Press, Princeton NJ (1954)
61. Wakker, P.P.: Explaining the characteristics of the power (CRRA) utility family. Health Econ. **17**(12), 1329–1344 (2008)
62. Wang, Y., Min, Q., Han, S.: Understanding the effects of trust and risk on individual behavior toward social media platforms: a meta-analysis of the empirical evidence. Comput. Hum. Behav. **56**, 34–44 (2016)
63. Weber, E.U., Johnson, E.J.: Decisions under uncertainty: psychological, economic, and neuroeconomic explanations of risk preference. In: Neuroeconomics, pp. 127–144. Academic Press (2009)
64. Woodford, M.: Prospect theory as efficient perceptual distortion. Am. Econ. Rev. **102**(3), 41–46 (2012)
65. Xie, D.: Power risk aversion utility function. Ann. Econ. Financ. **1**, 265–282 (2000)
66. Xie, J.: Discussion on the mechanism of irrational online shopping behavior—based on the perspective of mental accounting theory. Open J. Soc. Sci. **7**(05), 53 (2019)
67. Zhang, X., Ding, X., Ma, L.: The influences of information overload and social overload on intention to switch in social media. Behav. Inf. Technol. 1–14 (2020)

Concluding Remarks

The risk society that Ulrich Beck describes is essentially one where most critical risks are internally created rather than exogenous. It is precisely because of the sophistication of human endeavors, the advances in technology, and the growing complexity of economic activity that previously unknown threats and opportunities arise. Digital technologies have been instrumental in this transition. On the one hand they have acted as a magnifying glass for existing issues: for example, while misinformation has always existed, the scope and reach of social networks and the use of advanced targeting algorithms have weaponized it on an unprecedented scale, prodding some pundits to call it a weapon of math destruction. On the other hand, the digital world has created new phenomena such as merged reality and the almost infinite modularization of business, coupled with mass personalization of products. This has led to new classes of goods, services, assets, and transactions. The explosion of complexity also means an explosion of risk.

This is not to say that risk has never before been a crucial fact of economic and business life. Uncertainty has always existed, but the new digital reality calls for a more sophisticated understanding of risk which can be enabled by new analytic methods and approaches. The key insight from this book is that the analyst is now faced with emergent IT-driven risks that may have never materialized before. As usual, the data we use (from personal experience to extensive databases) is a thing from the past, while the risks we now face are a thing of the future. The greatest challenge is not merely to predict an outlandish realization but also to empower stakeholders to realize that such extreme events are actually plausible.

The analysis of very specific risks in this sense is much more difficult than the analysis of general ones. For instance, it may be nearly impossible to have predicted the emergence of a global online giant such as Amazon that simultaneously leverages advanced analytics, uses and sells cloud technologies and relies on AI-driven supply chain orchestration. Conversely, it is much easier to predict a large technology-driven shock on the revenue of traditional industries such as retail. The analysis may thus be better off by focusing on the more general level than trying to model highly

A. Gerunov, *Risk Analysis for the Digital Age*, Studies in Systems, Decision and Control 219, https://doi.org/10.1007/978-3-031-18100-9

uncertain events with extreme precision. A network-enabled approach could be used to pinpoint the drivers of this complexity and better understand possible influences.

The very approach to modeling risk may also have to be different. In a well-ordered traditional world, many relevant features of the environment are physically bounded. Those range from human characteristics such as intelligence or calories intake all the way into production features such as output, response, and delivery times. In those cases, the normal distribution may be an apt way to model the underlying phenomena. This need not be the case in the digital world. In many instances, it is a winner-takes-all world. For example, while there are many websites only very few of them take the lion's share of visits. Of the billions of apps in the digital ecosystems app stores, only a few attract the majority of users. Finally, incomes and revenue in the digital industry are also highly concentrated in the winners. Those new classes of digital phenomena necessitate a rethinking of how we model risks. Realizing that risk is the deviation from the expected, it must lurk in the tails of the statistical distribution. The analysis may need to use distributions with long or fat tails to appropriately capture the dynamics of digital uncertainty. To do this, the book proposes to switch away from the Gaussian distribution and experiment with distributions that better capture strange tail dynamics such as the lognormal or exponential one. This can be done in the framework of a Monte Carlo simulation that serves both as a generator of plausible outcomes, as well as an enabler of the analyst's imagination.

The final message of this book has to deal with the nature of risk. Its very essence carries well from "real life" to the digital world. Irrespective of where it originates, it is an objective part of the environment. While humans may fail to perceive it accurately, it still exists and does not disappear if we refuse to analyze and manage it. One may instead choose to transfer or pool it, to mitigate or even to accept it. Either way, the consequences of a possible realization will be felt by someone somewhere. The task of the analyst is thus not to pretend that risk has dissipated but rather to mitigate it to an acceptable level where an optimal decision can be made that balances the risk-adjusted benefits obtained with the costs incurred. It is the author's aspiration that this book has provided an introduction to some of the tools and ways of thinking that can effectively tackle this task.

Anton Gerunov

Sofia, January 2022